AF333154

Molecular Pathology and Genetics of Alport Syndrome

The generous support of Baxter Healthcare
is gratefully acknowledged

Contributions to Nephrology

Vol. 117

<table>
<tr><td>Series Editors</td><td>G.M. Berlyne, Brooklyn, N.Y.
S. Giovannetti, Pisa</td></tr>
<tr><td>Editorial Board</td><td>V. Bonomini, Bologna
J. Churg, New York, N.Y.
K.D.G. Edwards, New York, N.Y.
E.A. Friedman, Brooklyn, N.Y.
S.G. Massry, Los Angeles, Calif.
R.E. Rieselbach, Milwaukee, Wisc.
J. Traeger, Lyon</td></tr>
</table>

KARGER

Basel · Freiburg · Paris · London · New York ·
New Delhi · Bangkok · Singapore · Tokyo · Sydney

Molecular Pathology and Genetics of Alport Syndrome

Volume Editor   *K. Tryggvason,* Stockholm

48 figures and 19 tables, 1996

KARGER   Basel · Freiburg · Paris · London · New York · New Delhi · Bangkok · Singapore · Tokyo · Sydney

Contributions to Nephrology

Library of Congress Cataloging-in-Publication Data
Molecular pathology and genetics of Alport syndrome / volume editor, K. Tryggvason.
(Contributions to nephrology: vol. 117)
Includes bibliographical references and index.
1. Alport's syndrome – Molecular aspects. 2. Alport's syndrome – Genetic aspects. I. Tryggvason, K.
(Karl) II. Series: Contributions to nephrology; v. 117.
[DNLM: 1. Nephritis, Hereditary – genetics. 2. Nephritis, Hereditary – pathology. W1 CO778UN v. 117
1996 / WJ 353 M718 1996]
RF292.M65 1996
617.8'86 – dc20
ISBN 3–8055–6193–8 (hardcover: alk. paper)

Bibliographic Indices. This publication is listed in bibliographic services, including Current Contents® and Index Medicus.

Drug Dosage. The authors and the publisher have exerted every effort to ensure that drug selection and dosage set forth in this text are in accord with current recommendations and practice at the time of publication. However, in view of ongoing research, changes in government regulations, and the constant flow of information relating to drug therapy and drug reactions, the reader is urged to check the package insert for each drug for any change in indications and dosage and for added warnings and precautions. This is particularly important when the recommended agent is a new and/or infrequently employed drug.

© Copyright 1996 by S. Karger AG, P.O. Box, CH–4009 Basel (Switzerland)
Printed in Switzerland on acid-free paper by Thür AG Offsetdruck, Pratteln
ISBN 3–8055–6193–8

Contents

....

Preface

The elucidation of the pathogenesis of Alport syndrome is a typical example of the power of molecular biology. Professor Arthur Cecil Alport first described this hereditary nephritis characterized by hematuria and sensorineural deafness as a clinical syndrome in 1927, and electron microscopic indications of the presence of defects in the molecular structure of the glomerular basement membranes (GBM) were found at the end of the 1960s. However, it was not until 1990, following the discovery of the highly GBM-specific type IV collagen $\alpha 5$ chain, that the genetic basis for the X chromosome-linked type was determined. Mutations were observed in the gene encoding the type IV collagen $\alpha 5$ chain, and these findings were soon followed by the identification of mutations in the genes for the $\alpha 3$ and $\alpha 4$ chains. These three chains are now known to be highly specific to the GBM, where they are believed to have an important structural function. When this was written, well over 100 mutations in the three genes had been described in articles or congress abstracts. It is now possible to make exact DNA-based diagnoses of the disease, and current research is already aiming at gene therapy for Alport syndrome patients. Thus we have come a long way in Alport syndrome research in only a few years.

We have attempted in this book to compile state-of-the-art information on clinical findings and the pathology of Alport syndrome, with major emphasis on a detailed description of the molecular properties of the basement membrane, particularly the type IV collagen specific to it and the genes for this. The current status regarding mutations in type IV collagen genes is reviewed and methods for identifying such mutations are described. The chapters are written by investigators who have made fundamental contributions to the molecular genetics of type IV collagen and to the clarification of the pathogenesis of Alport syndrome, and I wish to thank all of them for their excellent work. I hope this book will be found useful by clinicians and basic scientists with interests in kidney diseases and basement membranes.

November 1995   *Karl Tryggvason*

Tryggvason K (ed): Molecular Pathology and Genetics of Alport Syndrome.
Contrib Nephrol. Basel, Karger, 1996, vol 117, pp 1–28

....

Alport Syndrome – Clinical Phenotypes, Incidence, and Pathology[1,2]

Martin C. Gregory[a], *Daniel A. Terreros*[b], *David F. Barker*[c],
Pamela N. Fain[d], *Joyce C. Denison*[c], *Curtis L. Atkin*[e]

Departments of [a]Medicine, [b]Pathology and Physiology, [c]Physiology,
[d]Medical Informatics, and [e]Biochemistry and Medicine, University of Utah,
Salt Lake City, Utah, USA

Hurst and Guthrie in the early part of this century studied a family with dominantly inherited hematuric nephritis. Alport's [1] contribution to the study of the condition was to recognize the profound hearing loss that accompanied nephritis in affected young males. This emphasis not only defined and publicized the disease, but unfortunately also hindered understanding by encouraging physicians and researchers to expect hearing loss as an obvious and integral part of the syndrome. It is now clear that the phenotypic expression of Alport syndrome (AS) is wide and includes cases with nephropathy alone [2, 3] as well as those with hearing loss. This realization is important, because, contrary to the initial belief, there may be as many cases of hereditary nephritis with apparently normal hearing as there are with hearing loss [4]. Furthermore, besides the kidney and the inner ear, other organs and cell types are involved to varying degrees. Abnormalities in the eye, skin, smooth muscle, platelets, and granulocytes are known so far.

[1] This work was supported by Public Health Services research grant No. MO1-RR00064 and NIH grant DK39497 and the Veterans Affairs Research Service.

[2] We thank Professor Jean-Pierre Grünfeld and Dr. Clifford Kashtan for comments on portions of the manuscript. We also thank Janis Perucca, Clark Gardner, Becky McLain, and Gregory Gilson for technical support with electron microscopy and histopathology.

Atkin et al. [3] divided AS into six phenotypes in 1988 (table 1). This classification has since been widely adopted, but it is now clear that the phenotypic expression is more complex. For example, there are families with undeniable AS with either granulocyte inclusions or diffuse leiomyomatosis. Autosomal recessive forms certainly occur. We doubt, however, that there is utility in expanding the phenotypic classification at present for either clinical purposes or research. From a clinical standpoint it suffices to know that: (a) the mode of transmission is heterogeneous, but commonly X-linked; (b) males are generally more severely affected than females in the X-linked form; (c) there is substantial heterogeneity in age of onset of both renal failure and hearing loss; (d) there may be associated defects of other organs, and (e) within a family a broadly similar phenotype can be expected in each affected member.

For research correlating mutations with clinical expression, it is important to characterize each item of potential heterogeneity in as many family members as possible. The best interim solution is to use the Atkin classification, with careful documentation of all relevant features.

As a cautionary note, there may be wide variance in the age of onset of the clinical features within a kindred. The mean age of onset of hearing loss or end-stage renal disease (ESRD) derived from several family members has more value than the time of appearance in an individual. In small kindreds the first classification should be considered provisional until more family members are evaluated. Researchers should ensure the genetic homogeneity of individuals grouped for descriptive purposes, or at least underscore the possible errors in assuming genetic similarity before this has been proven. Papers in which clinical features are collectively described without regard to potential genetic heterogeneity may be confusing.

1	Family history of nephritis or unexplained hematuria in a first-degree relative of the index case or in a male relative linked through any number of females
2	Persistent hematuria without evidence of an another possibly inherited nephropathy such as thin GBM disease, polycystic kidney disease or IgA nephropathy
3	Bilateral sensorineural hearing loss in the 2,000–8,000 Hz range; the hearing loss develops gradually, is not present in early infancy, and commonly presents before the age of 30 years
4	A mutation in COL4An (where n = 3, 4, or 5)
5	Immunohistochemical evidence of complete or partial lack of the Alport epitope in glomerular, or epidermal basement membranes, or both
6	Widespread GBM ultrastructural abnormalities, in particular thickening, thinning, and splitting
7	Ocular lesions including anterior lenticonus, posterior subcapsular cataract, posterior polymorphous dystrophy, and retinal flecks
8	Gradual progression to ESRD in the index case or at least two family members
9	Macrothrombocytopenia or granulocytic inclusions
10	Diffuse leiomyomatosis of esophagus, female genitalia, or both

Definition

How should AS be defined? Martin Bobrow at the First Alport Syndrome Workshop in 1990 in Finland emphasized the dichotomy between clinical and genetic bases of definition of AS. Physicians do not always have the information to make a genetic diagnosis, and need an operational definition as a basis for action. Conversely, from a genetic point of view, each family has a specific molecular lesion and therefore a molecular classification is rational. From a practical standpoint, we propose a clinical definition that we hope will eventually be replaced by a comprehensive molecular classification.

To establish the diagnosis of AS in a family for clinical or research purposes, at least four criteria (table 2) must be satisfied in closely related family members. Not all of the criteria need be found in the same person, but care must be taken when including distant relatives or few individuals whose sole manifestation is uninvestigated hematuria, ESRD, or hearing loss.

To diagnose AS in an individual, AS must be present in the family, the individual must be on the line of descent for the postulated mode of transmission and must satisfy one of criteria Nos. 2 through 10 for a diagnosis of probable AS and two criteria for a diagnosis of definite AS. To diagnose AS in an individual without a family history, at least four of the criteria must be present.

Clearly, some criteria are of greater value than others and a definition could give greater weight to them, but this would complicate matters. Most individuals satisfying a stringent criterion, such as Nos. 4 through 7, will also readily satisfy another such as No. 2. The definition satisfactorily discriminates the currently known forms of AS, including those that have arisen by new mutations, from thin glomerular basement membrane (GBM) disease and from other renal diseases.

As a practical point of differential diagnosis, any child who presents under the age of 6 years with visible hematuria should have a renal sonogram regardless of the family history because of the possibility of Wilms' tumor.

Phenotypes

Inheritance Pattern

The archetypal family showed dominant inheritance [1]. Classical inheritance patterns in nearly all convincing examples have been dominant, but molecular genetic analysis of a few sporadic cases or cases with affected siblings has shown clear evidence for autosomal recessive transmission [5–7] (see chapter 8).

After decades of discussion, inheritance in many kindreds was shown to be X-linked by classical pedigree analysis [8, 9], and by likelihood analysis [9, 10]. Strong genetic linkage of nephritis with restriction fragment length polymorphic markers (RFLPs) further strengthened the conclusion of X-linkage [11–16]. The COL4A5 gene at Xq22 that encodes the $\alpha5$ chain of type IV basement membrane collagen has been found defective in many cases of AS [17, 18], confirming the genetic mapping results. A summary of the mutations that have been associated with AS is given in chapter 8. The current status of COL4A5-linked genetic markers and their use in genetic linkage are discussed by Barker et al. in chapter 2 of this book.

Age at ESRD

Males in Alport kindreds develop ESRD at ages that vary widely from kindred to kindred, but that run broadly true within a kindred [19]. Schneider [20] recognized that age at ESRD is bimodal; in some families males develop ESRD in childhood or young adulthood, in others ESRD occurs in middle age. The concept of juvenile (mean age of ESRD in males <31 years) and adult (mean age of ESRD >31 years) kindreds was born and accepted [3, 19, 21], although it is undoubtedly an oversimplification. The age of ESRD is often rather broadly distributed within a kindred. Because an exceptional male in a juvenile kindred may not develop ESRD until over the age of 31, and because the converse is true for a few males in adult kindreds, one must be extremely careful in assigning phenotypes in small kindreds. Nevertheless, the distinction is worth retaining because juvenile and adult kindreds differ in several important respects: (a) hearing loss seems universal in juvenile kindreds but occurs overtly in only about half of adult kindreds; (b) ocular defects appear to be confined to juvenile kindreds, and (c) juvenile kindreds are more prone to develop anti-GBM disease after transplantation.

A diagnostic pitfall arises because pediatricians become familiar only with juvenile types of AS. A pediatrician seeing a boy with adult disease may have great difficulty in distinguishing him from one with familial thin basement membrane disease, even given an excellent family history, physical examination, and ultrastructural examination of a renal biopsy [22]. The concept that: 'Alport's syndrome could be differentiated from benign familial hematuria by history, physical examination, and routine laboratory testing' [23] is potentially fallible. A family history showing several males with hematuria who have reached advanced ages and none who have renal insufficiency is perhaps the best evidence that the condition is truly benign.

All cases of autosomal recessive AS described so far have been in children; females have been as severely affected as males [5–7]. Mutations have been found in the $\alpha3(IV)$ and $\alpha4(IV)$ chain in these sibships [5, 6].

Age of Onset and Rate of Progression of Sensorineural Hearing Loss

Almost all published information has been based on small numbers of genotypically heterogeneous patients, mostly with juvenile disease [24]. As a result, the conclusions cannot readily be generalized. Hearing loss is first noted at widely disparate ages [24–27], but tends to occur in early childhood and to be moderately severe to severe in juvenile kindreds. In the large adult-type Utah Kindred P, hearing loss roughly parallels the onset of renal insufficiency, whereas in an even larger kindred with a different mutation, hearing loss is not apparent [4, 28, 29]. Further analysis of the family without overt hearing loss has shown that among the few affected males who survive into their sixth decade, most develop hearing loss, appreciably later than the mean age of ESRD in this family [unpubl. observations], when it is difficult to distinguish from age- or noise-related hearing loss.

Mild hypofunction of the peripheral vestibular apparatus has been described but does not appear to be clinically significant [24, 26, 30, 31].

Ocular Lesions

Abnormalities of the cornea, lens, and retina occur in AS. They occur almost exclusively in juvenile AS, although exceptions occur [32]. Ocular lesions appear confined to families with hearing loss.

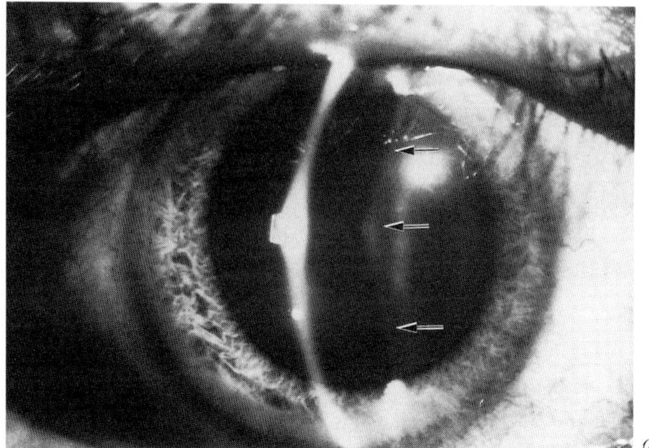
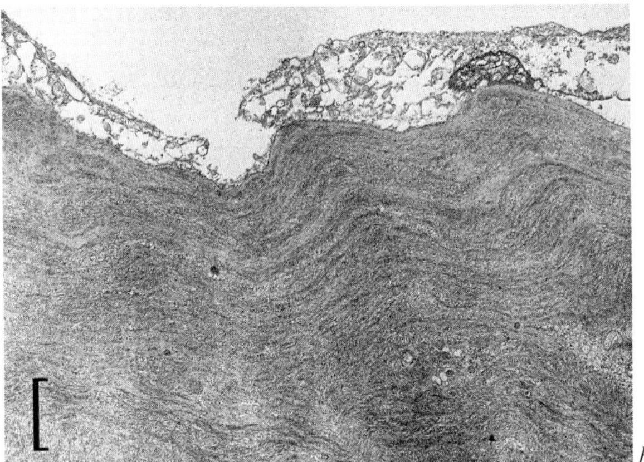

Fig. 1. a Slit-lamp photograph showing anterior lenticonus in the left eye of a patient with long-standing AS, juvenile type. The bright curved reflection is from the cornea. Laterally, the fainter reflection indicated by the arrows is from the anterior surface of the lens. The lenticonus is evident as a papilla indicated by the central arrow. Other reflections are illumination artifacts. *b* Electron microscopical features of the anterior lens capsule of the AS patient in figure 1a. Note the increased accumulation of fibrillar, collagen type, matrix.

Arcus cornealis [33] lacks specificity, but posterior polymorphous dystrophy appears to be a true association [34–36]. Sohar [30] first drew attention to ocular problems in AS. His description of spherophakia has been widely quoted, but it transpired that his cases actually had anterior lenticonus [37]. Anterior lenticonus (fig. 1a) has repeatedly been described in association with AS [32, 34, 38–45] and Nielsen [41] went so far as to suggest that 'lenticonus anterior exists exclusively as a part of Alport's syndrome'. Anterior lenticonus has been associated with X-linked disease [44, 46] and COL4A5 mutations [47]; moreover, AS and anterior lenticonus were fully expressed in an XO female [42]. Patients with autosomal recessive disease may also have anterior lenticonus [7]. Posterior lenticonus has rarely been described in cases of AS [48, 49]. Premature cataracts, usually anterior or posterior subcapsular, are common [32, 35, 45]. Small, yellow or white, perimacular retinal flecks are a reliable marker of AS [34–36, 43, 45, 50–52]. Rare but major complications include corneal abscesses and rupture of the anterior lens capsule.

Skin Immunochemistry

There are no clinically significant skin abnormalitiaes in AS, but there is an underlying abnormality of the basement membrane that can be shown immunochemically. Normal epidermal basement membrane reacts with antibodies against GBM, for example the familial nephritis serum obtained from an AS patient who developed anti-GBM antibodies after transplantation [53] and some other anti-GBM sera. Most hemizygous males with X-linked AS show no reaction with these antibodies, and heterozygous female carriers of X-linked AS show an interrupted pattern of reactivity [54]. This phenomenon is described in detail in chapter 7.

Leiomyomatosis

In several families and a number of sporadic cases with juvenile AS, nephritis and hearing loss have been closely associated with striking hypertrophy of the smooth muscle of the esophagus, female genitalia, or both [44, 46, 47, 55–61]. Families and individuals with these smooth muscle abnormalities have shown mutations that involve the 5' ends of both COL4A5 and COL4A6 [47]. While the inheritance is undoubtedly X-linked, these cases are perplexing in that the myocyte hypertrophy is particularly prominent at a fairly early age in heterozygous females. This intriguing association is discussed in detail by Antignac et al. in chapter 9.

Hematologic Abnormalities

Macrothrombocytopenia, the presence of abnormally few but exceptionally large blood platelets (fig. 2), has been associated with renal disease in at least 13 families [62–71] and 6 sporadic cases [27, 72–75]. Macrothrombocytopenia has consistently been inherited in an autosomal dominant fashion, whereas the nephritis is dominant but has not always segregated with the other manifestations. In many of the published pedigrees there is a distinct possibility that AS and a hematologic disease are being transmitted independently within the same kindred. Macrothrombocytopenia is a component of

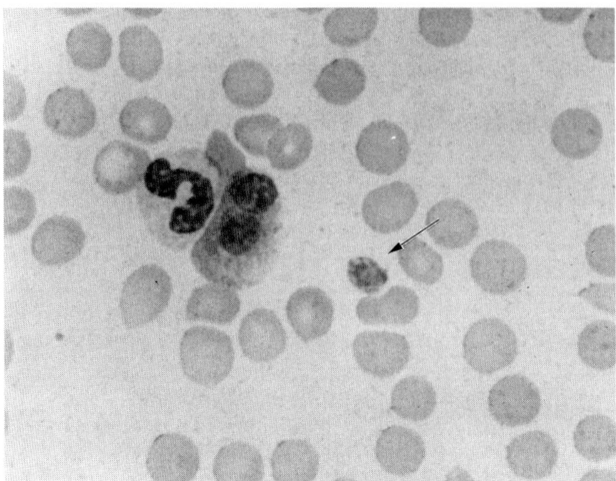

Fig. 2. Photomicrograph of a Wright-stained peripheral blood smear of an Epstein variant (type V) AS patient. A macroplatelet is present near the center, normal-appearing leukocytes are towards the upper left corner. The red cells have no abnormality. The platelets in this patient have variable sizes between 3 and 7 μm. The one in this field is about 5 μm (normal platelet size is 1–2 μm, erythrocytes are about 7 μm, and polymorphonuclear leukocytes 15 μm in diameter).

several congenital syndromes [76–78]. These are important because they may be found incidentally in a family or a single patient with nephropathy and cause diagnostic confusion (table 3). Nevertheless, the concurrence of nephritis and thrombocytopathy (Epstein syndrome) [62] has been noted so frequently that it appears clear that there is an autosomal dominant form of AS associated with macrothrombocytopenia and, in some families, granulocyte inclusions (see next section). The platelet count commonly ranges from 30,000 to 70,000. The platelets are often 5–15 μm in diameter and more spherical than normal. A mild bleeding tendency is usual, but severe bleeding after operations is not common.

In several families with macrothrombocytopenia, distinctive inclusions occur in circulating granulocytes. The first description was of May-Hegglin anomaly [66], a condition in which macrothrombocytopenia is associated with inclusion bodies in granulocytes and macrophages. A novel leukocytic inclusion was called the Fechtner inclusion after the man in whom it was first discovered [67]. The Fechtner inclusion is smaller and stains less well than the May-Hegglin inclusion [67]. This distinction is subtle. Because the Fechtner inclusion was not known at the time of Brivet's report [66], and because electron microscopy was not performed on these leukocytes, it is quite possible

Table 3. Inherited conditions associated with giant platelets and thrombocytopenia

Condition	Mode of inheritance	Associated nephropathy	Other associated conditions	References
Bernard-Soulier syndrome	AR	None	Severe bleeding tendency; prolonged bleeding time	76, 77
May-Hegglin anomaly	AD	Hematuria and ESRD	Prolonged bleeding time; Döhle bodies	66, 76
Gray platelet syndrome	AR	none	Prolonged bleeding time; absent α storage granules; marrow fibrosis	77
Epstein syndrome	AD?	Alport nephritis	Prolonged bleeding time; hearing loss	62
Fechtner syndrome	D	Alport nephritis	Fechtner inclusions; variable bleeding time; hearing loss; cataracts	67, 69
'Mediterranean' macrothrombo-cytopenia	Uncertain	None	Splenomegaly	139
'Montreal platelet syndrome'	AD	None	Prolonged bleeding time	76
Rafael platelet syndrome	AR?	None	Bleeding tendency; prolonged bleeding time	76, 77
Sebastian platelet syndrome	D	None	Fechtner inclusions; marginally prolonged bleeding time	78
Greaves syndrome	AD	None	Increased polyploidy of megakaryocytes	140

AR = Autosomal recessive; AD = autosomal dominant; D = dominant.

that the 'May-Hegglin' inclusions he described were actually Fechtner inclusions. Kindreds with macrothrombocytopenia and granulocyte inclusions have been found in France [66], the United States [67], Belgium [69], Israel [68], and Japan [70].

Charcot-Marie-Tooth Disease and AS

Several reports have linked chronic polyneuropathy with nephropathy, but the evidence that the renal lesion is that of AS is poor [79, 80].

Expression in Female Heterozygotes

In X-linked kindreds, expression in females is highly variable, but almost always much milder than in males. Hematuria is present in 88–93% of female gene carriers in adult kindreds [9, 15, 81]. Fewer female heterozygotes in juvenile kindreds display hematuria and the prevalence may be as low as 28% [9].

Severe renal failure is exceptional in female heterozygotes in X-linked kindreds, but ESRD does develop in up to 25%, generally at ages over 60 years [3]. Reports of renal failure in girls may represent newly characterized autosomal forms of the disease.

Expression of Autosomal AS

In rare dominant kindreds that are clearly unlinked to COL4A5, hearing loss and ESRD occur in adult life in both men and women [82]. Renal ultra-structure in a single biopsy showed widespread GBM thinning without clear evidence of lamellation or granulation. As discussed above, macrothrombocytopenia in AS has apparent autosomal dominant inheritance. In autosomal recessive sibships, females are as severely affected as males and hearing loss and ocular lesions are found in some families [5–7].

Incidence and Prevalence

There are no systematic surveys of the prevalence of AS in the general population. A gene frequency of 1:10,000 was estimated in Rhode Island [83] and we estimated the prevalence of AS in the Intermountain West of the United States at about 1:5,000 [10]. The prevalence in unselected renal biopsies was about 1.6% in Germany [84], although this figure doubles if 'hereditary progressive nephropathy without deafness' is included. AS accounts for 0.2–5% of those reaching ESRD in different series [85–90]. These figures are very imprecise and it is likely that many cases of X-linked hereditary nephritis without overt hearing loss have been overlooked [2, 4, 84].

Hematuria is present from birth in affected males, but as a result of selective attrition, the prevalence of juvenile types of AS falls with increasing ages of probands. Curves for the development of ESRD have been published previously [3]. Shaw and Kallen [83] estimated a mutation rate of 10 per million gametes with up to 18% of all newborns with the Alport genotype arising from new mutants. In Japan, 37% of children whose biopsies showed changes typical of AS did not have a family history and possibly were new mutants [91]. This figure may be an overestimate if children without a family history were more likely to undergo biopsy than those from a known kindred.

Pathogenesis

Understanding of the pathogenesis of AS was advanced by exploration of anti-GBM nephritis that occurred after transplantation. Transplantation is generally a very satisfactory form of treatment for ESRD in patients with AS, but in a few transplanted AS patients anti-GBM antibodies deposit on the GBM of the grafted kidney (fig. 3c) [92–94] and in even fewer [94] they destroy it by producing crescentic glomerulonephritis (fig. 3d) [87, 93–103]. Despite deposition of immunoglobulins on the GBM in up to a third of allografts, circulating anti-GBM antibodies are detectable much less often [92–94]. We speculate that a certain level of renal deposition of antibody can be tolerated without significant renal damage, but that once a certain degree of damage occurs, the greater exposure of 'foreign' basement membrane epitopes stimulates progressively more antibody production and increasingly severe injury. Most of these anti-GBM antibodies are of IgG class, but we have observed the concomitant development of IgA and IgM antibodies. In some patients, the antibodies react with the basement membrane of distal tubules.

Defects in the genes coding for the $\alpha5$ or $\alpha3$ chains of type IV collagen result in absent or imperfect formation of the heterotrimer of collagen IV that contains these chains and possibly the $\alpha4$ chain as well. The morphologic and functional consequences of AS result from the synthesis of abnormal collagen IV and its incorporation into basement membranes during embryogenesis, childhood, and adulthood. If absent during fetal life, immune tolerance to the $\alpha3$ chain does not develop, and anti-GBM antibodies form after this chain is introduced in the GBM of a transplanted kidney [103]. Although type IV collagen is more or less ubiquitous, clinically significant lesions of AS occur where it is most abundant, mainly in the kidneys, eyes, and inner ear. How mutations of basement membrane collagens may affect platelets and granulocytes is an enigma.

Pathology

Renal Pathology

Macroscopic Features
Little, if anything, is known about renal growth rates from birth to adolescence compared to controls. However, in both adult and juvenile types there is an accelerated reduction of renal size with aging [104]; not surprisingly, this process is more rapid in the juvenile variant. End-stage kidneys have markedly reduced combined weight (200 g or less) and can have small or large cortical

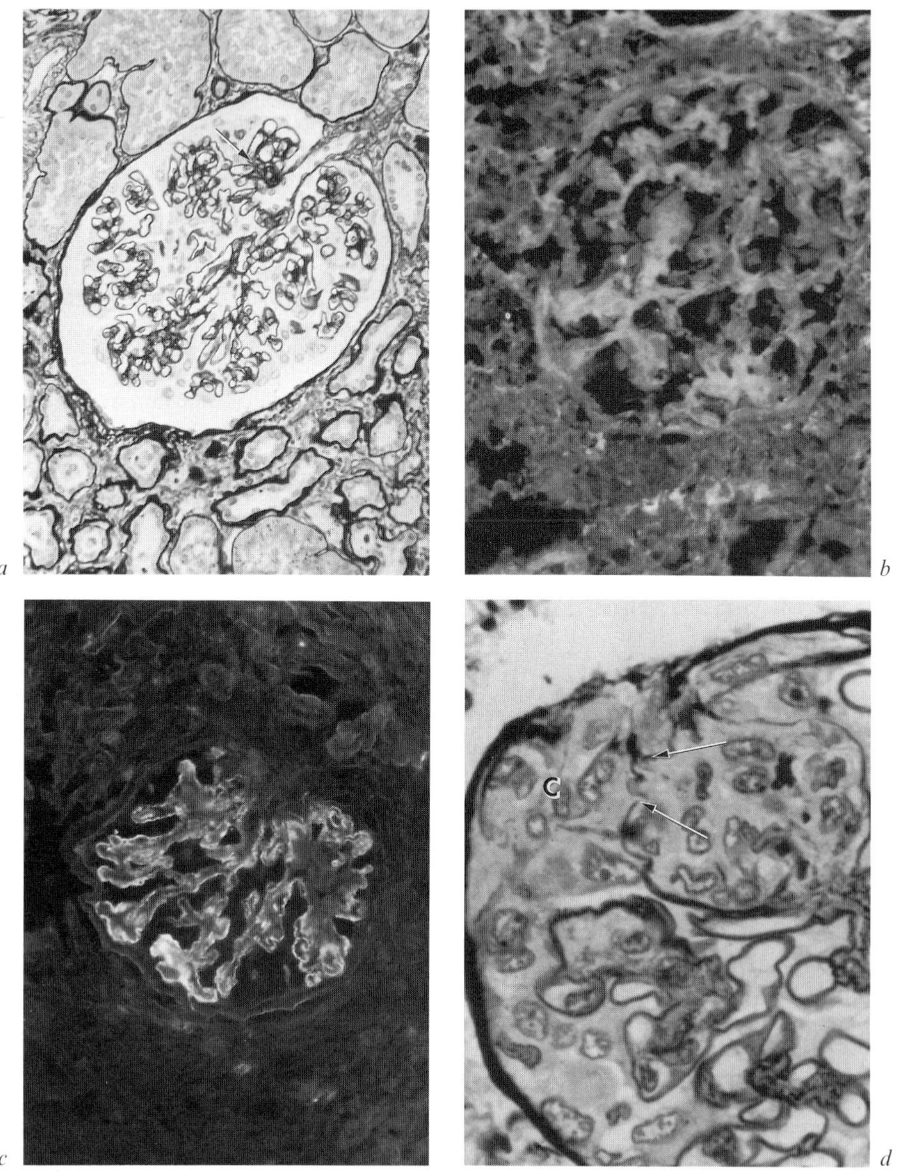

Fig. 3. a Photomicrograph of a silver methenamine-stained renal biopsy section from a patient with adult AS. The arrow indicates an area of segmental scarring. Otherwise, there are minimal abnormalities by light microscopy. *b* Photomicrograph of a renal biopsy section from a patient with adult AS pre-incubated with the serum from an Alport patient who had developed IgG anti-GBM antibodies following transplantation with a normal kidney. The section was stained for IgG and studied by indirect immunofluorescence. The basement membranes, in distinction to normal kidneys, did not react with anti-GBM antibody.

cysts filled with clear fluid. The capsule is markedly adherent and strips with difficulty leaving a finely granular uneven surface. The cortex is markedly thin while the medulla is relatively normal. There is a variable degree of lipid accumulation in the renal parenchyma.

Light Microscopy

The histopathological renal changes in AS depend on the specific phenotype and the stage of the disease at which the biopsy is obtained [105, 106]. There is no concrete evidence in either adult or juvenile AS about which histological changes are present in the pronephros, mesonephros, or embryonic metanephros. The persistence of fetal metanephric glomeruli into early childhood, at least in juvenile AS, suggests a renal maturation defect [107]. In the juvenile types, features consistent with AS can be seen in early childhood (fig. 5a). Uneven GBMs, minimal mesangial hypercellularity and scarring (fig. 3a) plus some tubular atrophy and intratubular red cell casts have been recorded at all ages in all types. Particularly in juvenile forms, there may be abundant cytoplasmic lipids in proximal tubular cells (fig. 5b) and in the interstitium (fig. 5c) [108]. Lipid accumulation in both epithelial cells and interstitial macrophages [109] is nonspecific. The lipid in these foam cells has been identified as cholesterol esters [J.I. O'Brien, unpubl. data]. Crescentic changes occur rarely [110]. Recently it has been claimed that AS can be recognized by light microscopy alone with a sensitivity of 72% and a specificity of 93% by looking for the triad of small capillary loops, poor staining with basement membrane stains, and presence of fetal-like glomeruli [111]. These features are rather subtle and such conclusions require independent confirmation. Biopsies of patients with adult type AS are rarely undertaken in childhood; when they are, only minimal abnormalities can be detected by light microscopy. After years of chronic glomerular capillary ruptures, the histological picture progresses through focal segmental scarring to diffuse glomerular sclerosis with secondary tubular and vascular atrophy. Perhaps because the glomerular capillaries at the base of the glomerular tuft are exposed to the highest pressure,

c Photomicrograph of an immunofluorescence stain for IgG on a renal graft biopsy section from a patient with juvenile AS after transplantation with a normal kidney. Linear IgG deposition is seen in this biopsy and corresponds to the development of anti-GBM antibodies. *d* Photomicrograph of a silver methenamine-stained renal biopsy section from the renal graft of a patient with juvenile AS after transplantation. Following the development of anti-GBM antibodies, the patient developed segmental capillary ruptures (arrows) with cellular crescents (C).

ruptures of the weakened glomerular capillaries are more frequent at the base. Glomerular scarring thus starts near the afferent arteriole [112] and gradually advances to involve the whole glomerular tuft. This process is likely to be accelerated by superimposition of overload nephropathy [113]. In patients in whom some degree of mesangial scarring and basal tuft scarring is present, the glomerular capillary walls are thickened [114]. This change results from hypertrophy of visceral epithelial cells with villous transformation and efface-ment of pedicels plus irregular thickening of the GBM [115]. Thickened capil-lary segments likely correspond to attempts at repair of partial capillary wall ruptures.

Immunofluorescence

Routine immunofluorescence studies are uniformly negative in the early stages of the renal injury. Nonspecific trapping of IgM, and to a lesser extent IgG and C3, can be seen later in areas of glomerular scarring. The overall importance of immunofluorescence studies in patients with familial hematuria lies in the need to rule out any form of inherited immune complex-mediated glomerulonephritis, such as familial IgA nephropathy, or superimposed im-mune complex-mediated disease. Focal segmental scarring and nonspecific IgM and complement deposition may erroneously suggest primary focal seg-mental glomerulosclerosis.

A specific and striking feature of GBMs in AS is their failure to bind anti-GBM antibodies (fig. 3b). This is true for anti-GBM sera occurring spontaneously in Goodpasture's syndrome [95, 116, 117], or in certain Alport males after transplantation [53, 54], as well as for $\alpha3(IV)$ anti-NC-1 monoclonal antibodies [118]. Complete lack of reactivity in males or interrupted loss in females is diagnostically helpful to separate AS from other basement mem-brane nephropathies. It is important to note that in a few families that amply satisfy criteria for AS, renal basement membranes react normally with these immune reactants [54]. These immune phenomena and their significance are dealt with in greater detail in chapter 7. Interestingly, in X-linked AS patients a second epitope is missing from the GBM [119]: amyloid P component is a lectin that binds to NC-1 chains. Its absence in AS can be demonstrated with monoclonal antibodies and presumably implies that defective NC-1 domains of type IV collagen do not facilitate accumulation of amyloid P in the GBM [119].

Electron Microscopy

Early in life, renal biopsies may show either no ultrastructural abnormali-ties [106, 107], or GBM thinning and splitting (fig. 5a) that may precede any clinical or laboratory signs other than hematuria. Eventually the GBMs become irregularly thinned and thickened with different degrees of splitting (fig. 4). In the most advanced cases the lamellation of the GBM creates uneven thickening and a 'basket weave' pattern (fig. 4a) [120] that likely represents ineffectual attempts at repair. The lamellae may be separated by numbers of ca. 50 nm diameter electron-dense granules [105, 108, 121]. In some patients, particularly heterozygous females, there is segmental hetero-geneity of the disease with segments that are completely normal alternating with areas of thinning and splitting. As a general observation, the ultrastruc-tural abnormalities are observed earlier and more dramatically in males than in females. In males, the thickening is usually not uniform and the GBM may be up to three times its normal thickness (fig. 4d). In children and young females, the most common observation is of diffuse or focal thinning of basement membranes to around 150 nm (fig. 4b) [27, 107, 122–124]. The normal thickness of GBM is 373 ± 42 nm in males, 326 ± 45 nm in females, and slightly less in children [125].

It is important to bear in mind that there are no pathognomonic transmis-sion electron microscopic changes in AS. Widespread lamellation of the GBM is highly characteristic for AS although it is not absolutely specific and may be seen focally and in conjunction with other lesions in a wide variety of nephropathies [22, 120, 121, 126]. Thickening of the GBM is a frequent feature of diabetic nephropathy [127]. Nevertheless, if thickening, thinning, la-mellation, and intramembranous granules are widespread in the absence of immune deposits or features or another disease, the diagnosis of AS is highly likely. Comparative ultrastructural studies of juvenile and adult AS as a func-tion of age are needed.

Inner Ear Pathology

Since the study of the auditory abnormalities in excised temporal bones requires significant postmortem mutilation, available information about the microscopic alterations in the inner ear is sparse. No abnormalities are evident in the tympanic membranes or ossicular chains [48, 128]. By light microscopic examination there is atrophy of the stria vascularis and vacuolization of the spiral ligament, and atrophy of the hair cells [26, 129, 130]. Others found no abnormalities [128]. In none of these reports was the specificity of the findings evaluated in reference to age-matched non-Alport controls with and without presbycusis or noise exposure.

Electron microscopy has been reported in only two cases [129]. Intercellu-lar edema in the region of the stria vascularis and a striking thickening and lamellation of the basal lamina of the strial vessels were found.

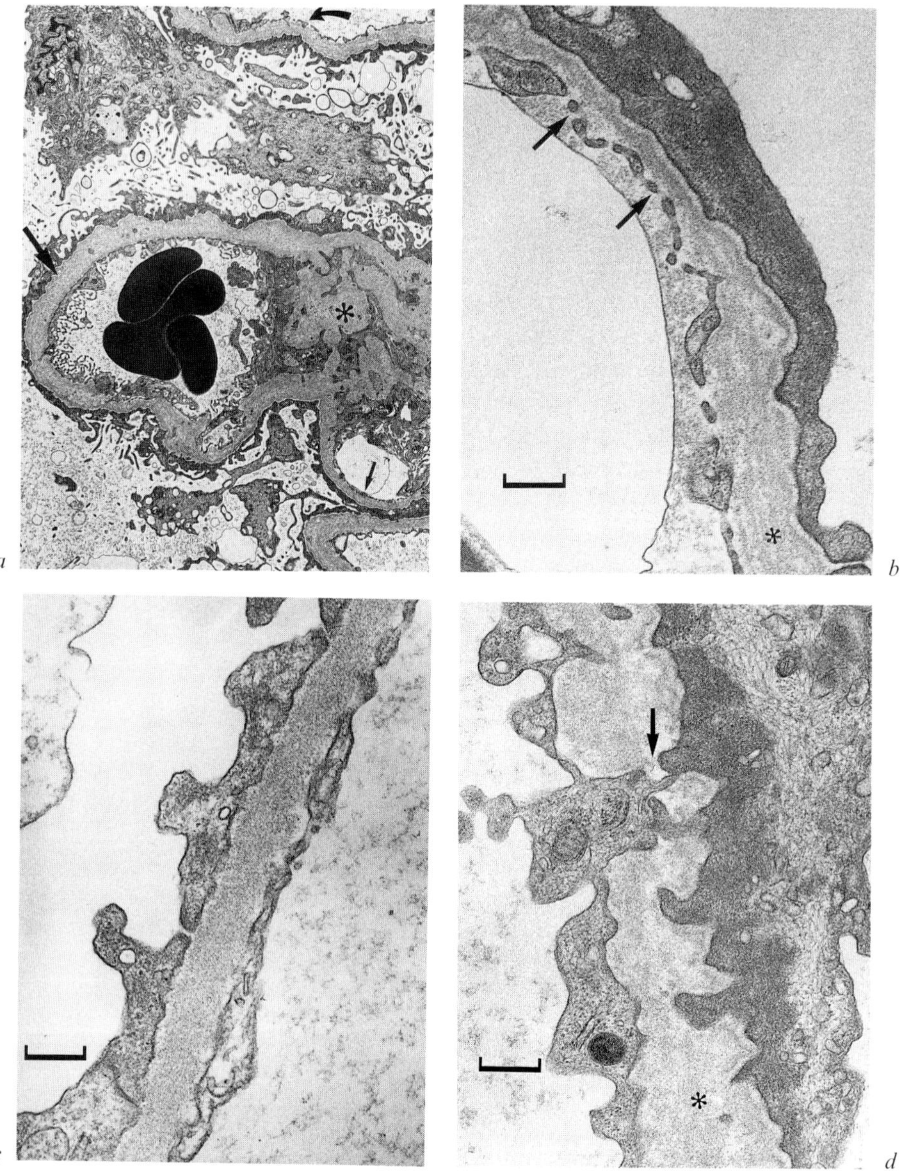

Fig. 4. *a* Electron microscopical glomerular features of a male patient with adult AS. There is extensive basket weaving and thickening of the GBM (large arrow), thinning and splitting (mid size arrow) and extensive lamellation and thickening (curved arrow). There is also fusion of pedicels and villous transformation of glomerular epithelial cells. The mesangium is segmentally scarred (asterisk). *b* Electron microscopical features of glomerular capillaries of a male patient with adult AS. There is focal thinning (arrows) and thickening with lamellation (asterisk) of the GBM. The endothelial cells (left of GBM) are not altered, the

The vestibule and semicircular canals appear normal by light microscopy, although there are abnormalities of the granular layer between the neuroepithelium and the otolithic membrane in the saccular maculae [128] and atrophy of the neuroepithelium of the cristae [26].

Ocular Pathology

Although less constant than sensorineural hearing loss, ocular abnormalities are fairly common in juvenile AS. Anterior conical deformity of the lens with protrusion into the anterior chamber of the eye (anterior lenticonus) is the best recognized and most diagnostically significant ocular abnormality [37, 39, 40, 131]. Pathologically the lens capsule varies in thickness [37, 39, 40, 131] with a more fibrillar appearance than normal (fig. 1b) and many partial capsular dehiscences [131]. Foveal granulations are frequently observed and are probably secondary to abnormalities of the retinal basement membrane or internal limiting membrane.

Hematopathology

Platelets and Megakaryocytes

In the rare families with macrothrombocytopenia, platelets show wide size disparity with large rounded platelets that may exceed 7 μm in greatest diameter [62–75]. Most platelets are spherical or elongated, rather than the normal discoid shape [62, 65, 67, 69]. In some accounts, the ultrastructure scarcely differs from normal [63, 68] whereas most authors note modest abnormalities of the subcellular structures. Granules, dense bodies, and sparse mitochondria are dispersed in the cytoplasm more irregularly than is usual [62, 67, 72]. Microtubules are generally normally distributed in a single plane in the platelets that are normally sized and discoid; in large spherical platelets,

visceral epithelial cells (right of GBM) have fused pedicels. × 9,200. *c* Electron microscopical features of glomerular capillaries of a normal 28-year-old female. The GBM is of uniform thickness (about 300 nm) and density. The endothelial cells are not altered and the pedicels have a normal architecture. × 9,200. *d* Electron microscopical features of a glomerular capillary of a male patient with adult AS. Compare the GBM with that in figure 4c at the same magnification. There is irregular thickening with lamellation (asterisk) and granularity of the GBM. There is also an almost complete rupture of the GBM as indicated by the arrow. An endothelial cell partially fills the gap. The visceral epithelial cells (right of GBM) have fused pedicels. × 9,200.

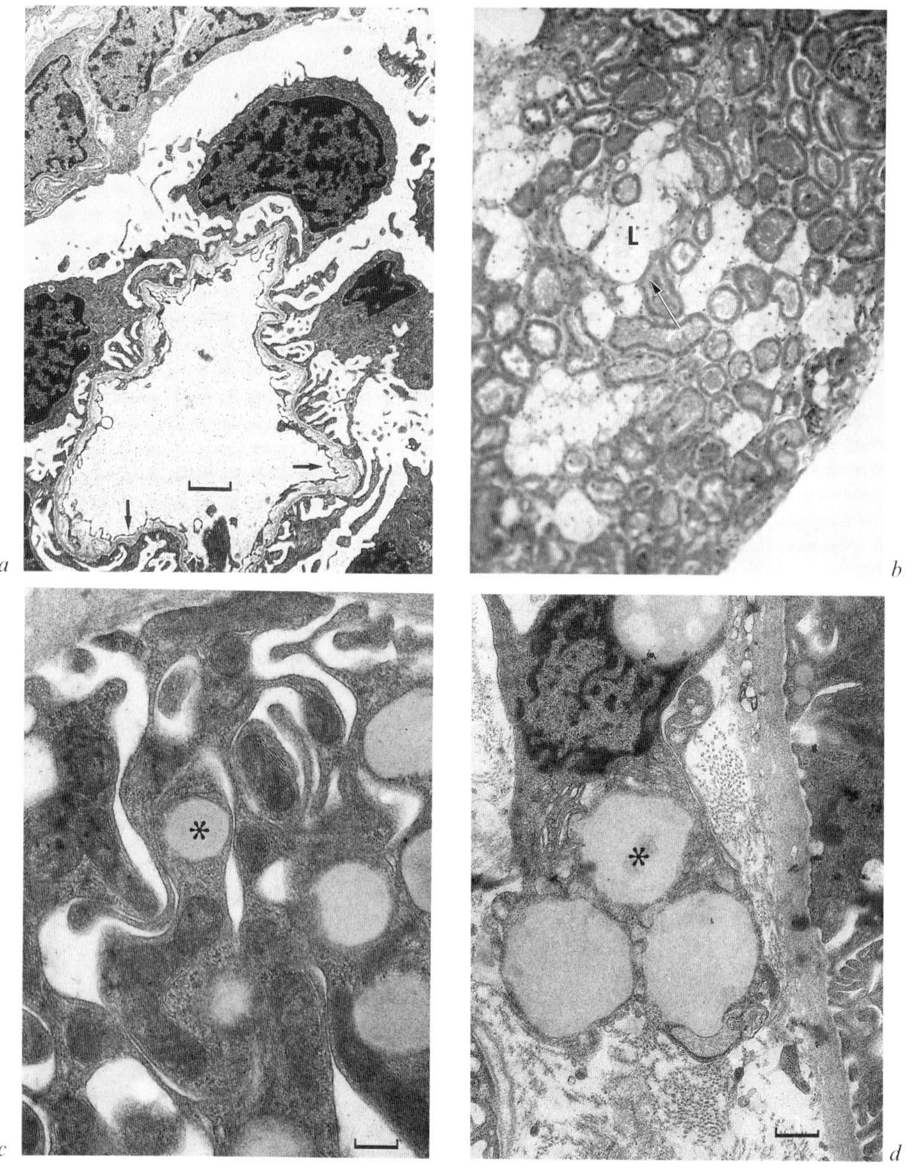

Fig. 5. a Electron microscopical glomerular features of a 1-year-old boy with juvenile AS. The GBMs are extensively lamellated and in some segments there is marked thickening (horizontal arrow), and in others the GBM appears almost normal (vertical arrow). × 2,300. *b* Photomicrograph of a trichrome stain of a renal biopsy section from a male with juvenile AS. This section of cortex shows abundant lipid (L) within proximal tubules (arrow). *c* Electron microscopical features of proximal renal tubular cells of the AS patient whose lipid-laden proximal renal tubular cells are depicted in figure 5b. At the top of the photograph

bands and bundles of microtubules lack this planar disposition [67, 69]. Elements of the surface-connected open canalicular system and channels of the dense tubular system are evenly spread within large platelets [67] and may intertwine to form striking maze-like areas of small tubules within the cytoplasm [62, 73]. Tubules within the maze are more uniform and of smaller diameter than normal smooth endoplasmic reticulum [62]. The surface-connected open canalicular system may be very prominent forming a sponge-like peripheral zone [71].

In the marrow, megakaryocytes and megakaryoblasts are plentiful [62, 64, 69]. Granules are distributed unevenly [62] with some aggregates of glycogen granules. Maze-like areas of small tubules within the megakaryocyte cytoplasm occur as in platelets [62]. Other abnormalities including a 'cartwheel' disposition of the demarcation membrane system have been described [69]. Megakaryocyte nuclei may have irregular profiles with numerous deep invaginations, and margination of nuclear chromatin [64], or condensation of the nucleus [69].

Leukocytes and Myelocytes

Granulocytes and monocytes in the May-Hegglin anomaly show one or more 2- to 5-μm long inclusions staining bright blue with May-Grünwald-Giemsa or Wright's stain. Most neutrophils and some eosinophils in Fechtner syndrome show one or more 1- to 2-μm pale blue irregular inclusions [67]. Fechtner inclusions are thus smaller and less densely stained than May-Hegglin inclusions. By electron microscopy, neutrophils appeared normal apart from the inclusions, which are small, irregular zones of cytoplasm free of granules, glycogen particles, and other organelles, but containing small clusters of single ribosomes [67]. Fechtner inclusions are less regular and lack the parallel 7- to 10-nm filaments characteristic of May-Hegglin inclusions.

Smooth Muscle Pathology

Multiple leiomyomata of the esophagus, stomach and vulva show no features to distinguish them from sporadic leiomyomata apart from the multicentric origin [55, 56]. This topic is covered in detail in chapter 9.

is the tubular basement membrane; there are abundant lipid droplets (asterisk) and mitochondria within the basolateral infoldings of this proximal renal tubular cell. × 9,200. *d* Electron microscopical features of a mononuclear interstitial cell in the kidney of the patient depicted in figure 5b and c. A lipid-laden monocyte (asterisk) is present in the interstitial space between a peritubular capillary (lower left corner) and a proximal tubule (right side of photograph). × 9,200.

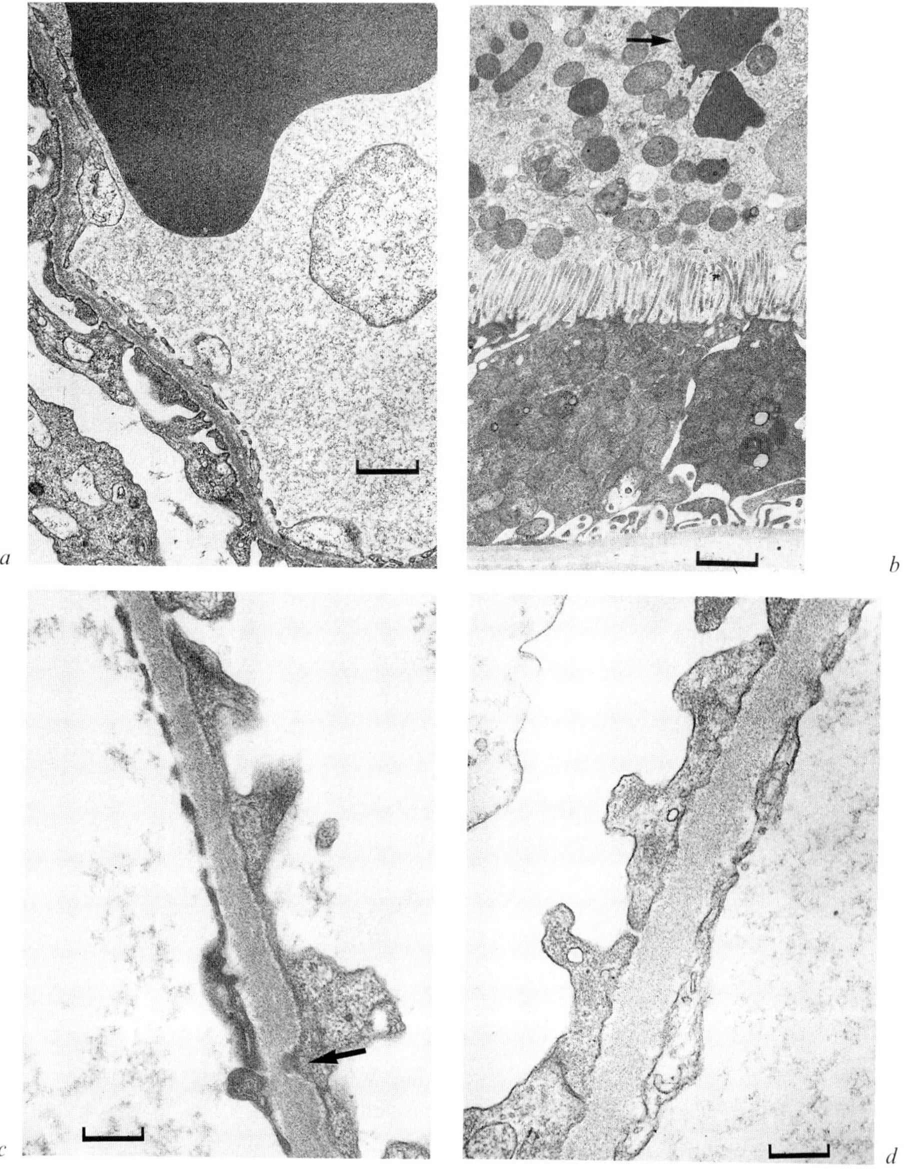

Fig. 6. a Electron microscopical glomerular features of a 27-year-old male patient with sporadic thin basement membrane disease. There is uniform thinning and density of the GBM without significant splitting. The endothelial and epithelial glomerular cells do not show significant abnormalities. A red cell is present within the capillary lumen. ×4,600. *b* Electron microscopical features observed in a proximal renal tubule of the 27-year-old male with thin membrane disease depicted in figure 6a. The tubular cells proper and tubular basement membranes do not have significant abnormality. There are dysmorphic red cells

Distinction from Familial Thin GBM Disease

Familial thin GBM disease, also know as 'benign familial hematuria', is transmitted as an autosomal dominant trait [23, 132–136]. Thin GBMs have been found in a number of individuals without a family history (fig. 6a) [137]. Subjects with thin basement membranes (fig. 6b, c) have lifelong microscopic hematuria (fig. 6b) but do not develop renal failure, and do not have ocular or inner ear abnormalities. Those few cases of deafness that have been reported [23] are likely coincidental [134] or may illustrate the difficulty distinguishing this condition from adult types of AS. Associated macrothrombocytopenia, granulocytic inclusions, or leiomyomatosis have not been observed in familial thin GBM disease.

Most importantly, although these patients have uniformly thin GBMs (fig. 6a, c), they possess the Alport epitope in glomerular and epidermal basement membranes and have no known mutations of COL4A5. Recognition of familial thin GBM disease is important to give patients and families an accurate prognosis and to avoid unnecessary investigation. Because a number of cases have developed renal failure [138] and because of the difficulty of distinguishing familial thin GBM disease from early cases of AS by electron microscopy alone, the family history should be as extended as possible, preferably showing several males with hematuria who have maintained a normal serum creatinine beyond the age of 60. Demonstration of normal reactivity of GBM with anti-GBM serum may help distinguish these two conditions, although some cases of AS do stain normally with anti-GBM serum. Whenever possible, evaluation for mutations in genes for type IV collagen should be undertaken. Demonstration of a mutation will establish a specific diagnosis although inability to find a mutation tilts the odds only slightly in favor of familial thin GBM disease. Taken together, these difficulties show how fallible the distinction between AS and familial thin GBM disease can be. Particularly if the proband is a child, family counseling about ultimate outcome should be guarded unless there is strong consistent information from an extensive family history including several elderly males, typical glomerular ultrastructure

(arrow) and cellular debris in the tubular lumen. ×4,600. *c* Electron microscopical features of a glomerular capillary of the same patient depicted in figure 6a. The GBM thickness is about half that of the normal GBM in figure 6d. The GBM is uniformly thin without density abnormalities or lamellation. The endothelial and epithelial cells do not have diagnostic abnormalities. Occasionally as depicted by the arrow, there are partial GBM ruptures, in this case filled by an epithelial cell. ×9,200. *d* Electron microscopical features of a normal glomerular capillary (same photograph as figure 4c) for comparison of GBM with figure 6c. ×9,200.

in at least one family member, and positive staining of glomerular capillary walls with an appropriate antiserum, or evidence of lack of a relevant mutation in basement membrane collagen genes.

In summary: AS is genetically and phenotypically heterogeneous. Most cases are X-linked dominant, but autosomal dominant and recessive forms also occur. A keen awareness of phenotypic diversity is essential for research, diagnosis, and for genetic counseling. Two main subgroups exist, based on the age of onset of ESRD. In juvenile kindreds, the mean age of ESRD in males is about 18 years; in adult kindreds it is about 37 years. Thirty-one years is a good dividing line between the two types. Juvenile kindreds are small, perhaps because of poor reproductive fitness, but numerous, presumably reflecting high mutation and extinction rates. Juvenile kindreds generally have hearing impairment, and tend to have ocular defects. Interestingly, anti-GBM nephritis following transplantation seems to occur mainly in juvenile kindreds. Adult kindreds include few, but large families. Adult disease is characterized by low mutation and extinction rates and by normal reproductive fitness. In adult kindreds, hearing loss may be delayed and not apparent until well after the age of ESRD. These cases are quite common, but are readily overlooked because both hearing loss and renal failure occur relatively late in life.

References

1   Alport AC: Hereditary familial congenital haemorrhagic nephritis. Br Med J 1927;i:504–506.
2   Dockhorn RJ: Hereditary nephropathy without deafness. Am J Dis Child 1967;114:135–138.
3   Atkin CL, Gregory MC, Border WA: Alport Syndrome. Diseases of the Kidney. Boston, Little, Brown, 1988, pp 617–641.
4   Gregory MC, Skinner B, Atkin CL, Barker DF: A novel mutation in COL4A5 relates three families with type IV Alport syndrome. J Am Soc Nephrol 1991;2:254.
5   Lemmink HH, Mochizuki T, van den Heuvel LPWT, Schröder CH, Barrientos A, Monnens LAH, et al: Mutations in the type IV collagen α3(COL4A3) gene in autosomal recessive Alport syndrome. Hum Molec Genet 1994;8:1269–1273.
6   Mochizuki T, Lemmink HH, Mariyama M, Antignac C, Gubler M-C, Pirson Y, et al: Identification of mutations in the α3(IV) and α4(IV) collagen genes in autosomal recessive Alport syndrome. Nature Genetics 1994;8:77–81.
7   Knebelmann B, Benessy F, Buemi M, Grünfeld JP, Gubler MC, Antignac C: Autosomal recessive inheritance pattern in Alport syndrome. J Am Soc Nephrol 1993;4:263–263.
8   O'Neill WMJ, Atkin CL, Bloomer HA: Hereditary nephritis: A re-examination of its clinical and genetic features. Ann Intern Med 1978;88:176–182.
9   Hasstedt SJ, Atkin CL, San Juan AC Jr: Genetic heterogeneity among kindreds with Alport syndrome. Am J Hum Genet 1986;38:940–953.
10 Hasstedt SJ, Atkin CL: X-linked inheritance of Alport syndrome: Family P revisited. Am J Hum Genet 1983;35:1241–1251.
11 Menlove L, Aldridge J, Schwartz C, Atkin C, Hasstedt S, Kunkel L, et al: Linkage between Alport syndrome-like hereditary nephritis and X-linked RFLPs. Am J Hum Genet 1984;36:146S.
12 Atkin CL, Hasstedt SJ, Menlove L, Cannon L, Kirschner N, Schwartz C, et al: Mapping of Alport syndrome to the long arm of the X chromosome. Am J Hum Genet 1988;42:249–255.

13 Szpiro-Tapia S, Bobrie G, Guilloud BM, Heuertz S, Julier C, Frezal J, et al: Linkage studies in X-linked Alport's syndrome. Hum Genet 1988;81:85–87.

14 Brunner H, Schröder C, Monnens L, Veerkamp J, Ropers HH: Alport's syndrome: Localization of the X-chromosomal gene and consequences for future investigations. Contrib Nephrol. Basel, Karger, 1988, vol 67, pp 200–205.

15 Barker DF, Fain PR, Goldgar DE, Dietz-Band JN, Turco AE, Kashtan CE, et al: High-density genetic and physical mapping of DNA markers near the X-linked Alport syndrome locus: Definition and use of flanking polymorphic markers. Hum Genet 1991;88:189–194.

16 Vetrie D, Flinter F, Bobrow M, Harris A: Long-range mapping of the gene for the human a5(IV) collagen chain at Xq22-q23. Genomics 1992;12:130–138.

17 Barker DF, Hostikka SL, Zhou J, Chow LT, Oliphant AR, Gerken SC, et al: Identification of mutations in the COL4A5 collagen gene in Alport syndrome. Science 1990;248:1224–1227.

18 Tryggvason K, Zhou J, Hostikka SL, Shows TB: Molecular genetics of Alport syndrome. Kidney Int 1993;43:38–44.

19 Tishler PV, Rosner B: The genetics of the Alport syndrome. Birth Defects 1974;10:93–99.

20 Schneider RG: Congenital hereditary nephritis with nerve deafness. N Y State J Med 1963;15: 2644–2648.

21 Grünfeld JP, Bois EP, Hinglais N: Progressive and nonprogressive hereditary chronic nephritis. Kidney Int 1973;4:216–228.

22 Hill GS, Jenis EH, Goodloe SJ: The nonspecificity of the ultrastructure lesion in hereditary nephritis: With additional observations on benign familial hematuria. Lab Invest 1974;31:516–532.

23 Blumenthal SS, Fritsche C, Lemann JJ: Establishing the diagnosis of benign familial hematuria. The importance of examining the urine sediment of family members. JAMA 1988;259:2263–2266.

24 Gleeson MJ: Alport's syndrome: Audiological manifestations and implications. J Laryngol Otol 1984;98:449–465.

25 Winter LE, Cram BM, Banovetz JD: Hearing loss in hereditary renal disease. Arch Otolaryngol 1968;88:238–241.

26 Miller GW, Joseph DJ, Cozad RL, McCabe BF: Alport's syndrome. Arch Otolaryngol 1970;92: 419–432.

27 Gubler M, Levy M, Broyer M, Naizot C, Gonzales G, Perrin D, et al: Alport's syndrome. A report of 58 cases and a review of the literature. Am J Med 1981;70:493–505.

28 Gregory MC, Atkin CL: Alport syndrome; in Schrier RW, Gottschalk CW (eds): Diseases of the Kidney, ed 5. Boston, Little, Brown, 1993, pp 571–591.

29 Wester DC, Atkin CL, Gregory MC: Alport syndrome: clinical update. J Am Acad Audiol 1995; 6:73–79.

30 Sohar E: Renal disease, inner ear deafness, and ocular changes. Arch Intern Med 1956;97:627–630.

31 Celis-Blaubach A, Garcia-Zozaya JL, Perez-Requejo JL, Brasse K: Vestibular disorders in Alport's syndrome. J Laryngol Otol 1974;88:663–674.

32 Chance JK, Stanley JA: Alport's syndrome: Case report and review of ocular manifestations. Ann Ophthalmol 1977;9:1527–1530.

33 Chavis RM, Groshong T: Corneal arcus in Alport's syndrome. Am J Ophthalmol 1973;75:793–794.

34 Sabates R, Krachmer JH, Weingeist TA: Ocular findings in Alport's syndrome. Ophthalmologica 1983;186:204–210.

35 Thompson SM, Deady JP, Willshaw HE, White RH: Ocular signs in Alport's syndrome. Eye 1987; 1:146–153.

36 Teekhasaenee C, Nimmanit S, Wutthiphan S, Vareesangthip K, Laohapand T, Malasitr P, et al: Posterior polymorphous dystrophy and Alport syndrome. Ophthalmology 1991;98:1207–1215.

37 Arenberg IK, Dodson VN, Falls HF, Stern SD: Alport's syndrome: Reevaluation of the associated ocular abnormalities and report of a family study. J Pediatr Ophthalmol 1967;4:21–32.

38 Arnott EJ, Crawfurd MD, Toghill PJ: Anterior lenticonus and Alport syndrome. Br J Ophthalmol 1966;50:390–403.

39 Brownell RD, Wolter JR: Anterior lenticonus in familial hemorrhagic nephritis. Arch Ophthalmol 1964;71:481–483.

40   Faggioni R, Scouras J, Streiff EB: Alport's syndrome. Clinicopathological considerations. Ophthalmologica 1972;165:1–14.

41   Nielsen CE: Lenticonus anterior and Alport's syndrome. Acta Ophthalmol (Copenh) 1978;56:518–530.

42   Kapoor S, Dasgupta J: Chromosomal anomaly in a female patient with anterior lenticonus. Ophthalmologica 1979;179:271–275.

43   Govan JA: Ocular manifestations of Alport's syndrome: A hereditary disorder of basement membranes? Br J Ophthalmol 1983;67:493–503.

44   McCartney PJ, McGuinness R: Alport's syndrome and the eye. Aust N Z J Ophthalmol 1989;17:165–168.

45   Jacobs M, Jeffrey B, Kriss A, Taylor D, Sa G, Barratt M: Ophthalmologic assessment of young patients with Alport syndrome. Ophthalmology 1992;99:1039–1044.

46   Legius E, Proesmans W, Van DB, Geboes K, Lerut T, Eggermont E: Muscular hypertrophy of the oesophagus and 'Alport-like' glomerular lesions in a boy. Eur J Pediatr 1990;149:623–627.

47   Antignac C, Zhou J, Sanak M, Cochat P, Roussel B, Deschenes G, et al: Alport syndrome and diffuse leiomyomatosis: Deletions in the 5' end of the COL4A5 collagen gene. Kidney Int 1992;42:1178–1183.

48   Crawfurd MD'A, Toghill PJ: Alport's syndrome of hereditary nephritis and deafness. Q J Med 1968;37:563–576.

49   Bhatnagar R, Kumar A, Pakrasi S: Alport's syndrome – Ocular manifestations and unusual features. Acta Ophthalmol (Copenh) 1990;68:347–349.

50   Perrin D, Jungers P, Grünfeld JP, Delons S, Noel LH, Zenatti C: Perimacular changes in Alport's syndrome. Clin Nephrol 1980;13:163–167.

51   Zylbermann R, Silverstone BZ, Brandes E, Drukker A: Retinal lesions in Alport's syndrome. J Pediatr Ophthalmol Strabismus 1980;17:255–260.

52   Gelisken O, Hendrikse F, Schroder CH, Berden JH: Retinal abnormalities in Alport's syndrome. Acta Ophthalmol (Copenh) 1988;66:713–717.

53   Kashtan C, Fish AJ, Kleppel M, Yoshioka K, Michael AF: Nephritogenic antigen determinants in epidermal and renal basement membranes of kindreds with Alport-type familial nephritis. J Clin Invest 1986;78:1035–1044.

54   Kashtan CE, Atkin CL, Gregory MC, Michael AF: Identification of variant Alport phenotypes using an Alport-specific antibody probe. Kidney Int 1989;36:669–674.

55   Johnston JB, Clagett OT, McDonald JR: Smooth-muscle tumors of the oesophagus. Thorax 1953;8:251–265.

56   Garcia-Torres R, Guarner V: Leiomyomatosis of the esophagus, tracheobronchi and genitals associated with Alport-type hereditary nephropathy: A new syndrome. Rev Gastroenterol Méx 1983;48:163–170.

57   Roussel B, Birembaut P, Gaillard D, Puchelle JC, D'Albignac G, Pennaforte F, et al: Familial esophageal leiomyomatosis associated with Alport's syndrome in a 9-year-old boy. Helv Paediatr Acta 1986;41:359–368.

58   Cochat P, Guibaud P, Garcia TR, Roussel B, Guarner V, Larbre F: Diffuse leiomyomatosis in Alport syndrome. J Pediatr 1988;113:339–343.

59   Wittig BM, Treichel U, Kohler H, Rumpelt HJ HJ, Meyer zum Buschenfelde KH: Leiomyomatose des Gastrointestinal – und Urogenitaltrakts in Kombination mit hereditärer Nephritis. Med Klin 1990;85(suppl 1):122–124.

60   Rabushka LS, Fishman EK, Kuhlman JE, Hruban RH: Diffuse esophageal leiomyomatosis in a patient with Alport syndrome: CT demonstration. Radiology 1991;179:176–178.

61   Lonsdale RN, Roberts PF, Vaughan R, Thiru S: Familial oesophageal leiomyomatosis and nephropathy. Histopathology 1992;20:127–133.

62   Epstein CJ, Sahud MA, Piel CF, Goodman JR, Bernfield MR, Kushner JH, Ablin AR: Hereditary macrothrombocytopathia, nephritis and deafness. Am J Med 1972;52:299–310.

63   Eckstein JD, Filip DJ, Watts JC: Hereditary thrombocytopenia, deafness, and renal disease. Ann Intern Med 1975;82:639–645.

64   Parsa KP, Lee DB, Zamboni L, Glassock RJ: Hereditary nephritis, deafness and abnormal thrombopoiesis. Study of a new kindred. Am J Med 1976;60:665–672.

65   Clare NM, Montiel MM, Lifschitz MD, Bannayan GA: Alport's syndrome associated with macrothrombopathic thrombocytopenia. Am J Clin Pathol 1979;72:111–117.

66   Brivet F, Girot R, Barbanel C, Gazengel C, Maier M, Crosnier J: Hereditary nephritis associated with May-Hegglin anomaly. Nephron 1981;29:59–62.

67   Peterson LC, Rao KV, Crosson JT, White JG: Fechtner syndrome – A variant of Alport's syndrome with leukocyte inclusions and macrothrombocytopenia. Blood 1985;65:397–406.

68   Gershoni-Baruch R, Baruch Y, Viener A, Lichtig C: Fechtner syndrome: Clinical and genetic aspects. Am J Med Genet 1988;31:357–367.

69   Heynen MJ, Blockmans D, Verwilghen RL, Vermylen J: Congenital macrothrombocytopenia, leucocyte inclusions, deafness and proteinuria: functional and electron microscopic observations on platelets and megakaryocytes. Br J Haematol 1988;70:441–448.

70   Takai K, Sanada M, Hattori A, Koike T, Shibata A: Fechtner syndrome: Report of two families and review of the literature on the related disorders. Nippon Ketsueki Gakkai Zasshi 1989;52: 644–654.

71   Túri S, Kóbor J, Erdös A, Bodrogi T, Virág I, Ormos J: Hereditary nephritis, platelet disorders and deafness – Epstein's syndrome. Pediatr Nephrol 1992;6:38–43.

72   Bernheim SJ, Dechavanne M, Bryon PA, Lagarde M, Colon S, Pozet N, et al: Thrombocytopenia, macrothrombocytopathia, nephritis and deafness. Am J Med 1976;61:145–150.

73   Hansen MS, Behnke O, Pedersen NT, Videbaek A: Megathrombocytopenia associated with glomerulonephritis, deafness, and aortic cystic media necrosis. Scand J Haematol 1978;21:197–205.

74   Thomas HS, Bauer JH: Hereditary nephritis, deafness and thrombocytopenia. Case report and review. Mo Med 1984;81:305–311.

75   Standen GR, Saunders J, Michael J, Bloom AL: Epstein's syndrome: Case report and survey of the literature. Postgrad Med J 1987;63:573–575.

76   Milton JG, Hutton RA, Tuddenham EGD, Frojmovic MM: Platelet size and shape in hereditary giant platelet syndromes on blood smear and in suspension: Evidence for two types of abnormalities. J Lab Clin Med 1985;106:326–335.

77   Nichols WL, Kaese SE, Gastineau DA, Otteman LA, Bowie EJW: Bernard-Soulier syndrome: Whole blood diagnostic assays of platelets. Mayo Clin Proc 1989;64:522–530.

78   Greinacher A, Mueller-Eckhardt C: Hereditary types of thrombocytopenia with giant platelets and inclusion bodies in the leukocytes. Blut 1990;60:53–60.

79   Lemieux G, Neemeh JA: Charcot-Marie-Tooth disease and nephritis. Can Med Assoc J 1967;97: 1193–1198.

80   Paul MD, Fernandez D, Pryse-Phillips W, Gault MH: Charcot-Marie-Tooth disease and nephropathy in a mother and daughter with a review of the literature. Nephron 1990;54:80–85.

81   Tishler PV: Healthy female carriers of a gene for the Alport syndrome: Importance for genetic counseling. Clin Genet 1979;16:291–294.

82   Gregory MC, Fain PN, Donaldson C, Barker DF, Atkin CL: Small-kindred linkage analysis reveals genetic heterogeneity in Alport syndrome. J Am Soc Nephrol 1991;2:254.

83   Shaw RF, Kallen RJ: Population genetics of Alport's syndrome. Hypothesis of abnormal segregation and the necessary existence of mutation. Nephron 1976;16:427–432.

84   Waldherr R: Familial glomerular disease. Contrib Nephrol. Basel, Karger, 1982, vol 33, pp 104–121.

85   US Renal Data System. National Institutes of Health, National Institute of Diabetes and Digestive and Kidney Diseases, Bethesda, Md. USRDS 1993 Annual Data Report 1993, pp 24–24.

86   Wing AJ, Broyer M, Brunner FP, et al: Combined report on regular dialysis and transplantation in Europe, XIII, 1982. Proc EDTA 1983;20:5.

87   Milliner DS, Pierides AM, Holley KE: Renal transplantation in Alport's syndrome: Anti-glomerular basement membrane glomerulonephritis in the allograft. Mayo Clin Proc 1982;57:35–43.

88   Broyer M: Chronic renal failure. Pediatr Nephrol 1974;358–394.

89   Chantler C: Familial renal disease; in Weatherall DJ, Ledingham JGG, Warrell DA (eds): Oxford Textbook of Medicine. Oxford, Oxford Medical Publications, 1987, pp 18.81–18.87.

90 Green A, Allos M, Donohoe J, Carmody M, Walshe J: Prevalence of hereditary renal disease. Ir Med J 1990;83:11–13.

91 Yoshikawa N, Matsuyama S, Ito H, Hahikano H, Matsuo T: Nonfamilial hematuria associated with glomerular basement membrane alterations characteristic of hereditary nephritis: Comparison with hereditary nephritis. J Pediatr 1987;111:519–524.

92 Quérin S, Noël LH, Grünfeld JP, Droz D, Mahieu P, Berger J, et al: Linear glomerular IgG fixation in renal allografts: Incidence and significance in Alport's syndrome. Clin Nephrol 1986;25:134–140.

93 Göbel J, Olbricht CJ, Offner G, Helmchen U, Repp H, Koch KM, et al: Kidney transplantation in Alport's syndrome: Long-term outcome and allograft anti-GBM nephritis. Clin Nephrol 1992;38:299–304.

94 Peten E, Pirson Y, Cosyns J, Squifflet J, Alexandre GPJ, Noël L, et al: Outcome of thirty patients with Alport's syndrome after renal transplantation. Transplantation 1991;52:823–826.

95 McCoy RC, Johnson HK, Stone WJ, Wilson CB: Absence of nephritogenic GBM antigen(s) in some patients with hereditary nephritis. Kidney Int 1982;21:642–652.

96 Teruel JL, Liano F, Mampaso F, Moreno J, Serrano A, Quereda C, et al: Allograft antiglomerular basement membrane glomerulonephritis in a patient with Alport's syndrome. Nephron 1987;46:43–44.

97 Fleming SJ, Savage CO, McWilliam LJ, Pickering SJ, Ralston AJ, Johnson RW, et al: Anti-glomerular basement membrane antibody-mediated nephritis complicating transplantation in a patient with Alport's syndrome. Transplantation 1988;46:857–859

98 Shah B, First MR, Mendoza NC, Clyne DH, Alexander JW, Weiss MA: Alport's syndrome: Risk of glomerulonephritis induced by antiglomerular-basement-membrane antibody after renal transplantation. Nephron 1988;50:34–38.

99 Van den Heuvel LP, Schröder CH, Savage CO, Menzel D, Assmann KJ, Monnens LA, et al: The development of anti-glomerular basement membrane nephritis in two children with Alport's syndrome after renal transplantation: Characterization of the antibody target. Pediatr Nephrol 1989;3:406–413.

100 Kashtan CE, Butkowski RJ, Kleppel MM, First MR, Michael AF: Post-transplant anti-glomerular basement membrane nephritis in related males with Alport syndrome. J Lab Clin Med 1990;116:508–515.

101 Goldman M, Depierreux M, De Pauw L, Vereerstraeten P, Kinnaert P, Noel LH, et al: Failure of two subsequent renal grafts by anti-GBM glomerulonephritis in Alport's syndrome: Case report and review of the literature. Transplant Int 1990;3:82–85.

102 Rassoul Z, Al Khader AA, Al Sulaiman M, Dhar JM, Coode P: Recurrent allograft antiglomerular basement membrane glomerulonephritis in a patient with Alport's syndrome. Am J Nephrol 1990;10:73–76.

103 Hudson BG, Kalluri R, Gunwar S, Weber M, Ballester F, Hudson JK, et al: The pathogenesis of Alport syndrome involves type IV collagen molecules containing the a3(IV) chain: Evidence from anti-GBM nephritis after renal transplantation. Kidney Int 1992;42:179–187.

104 Krickstein HI, Gloor FJ, Balogh KJ: Renal pathology in hereditary nephritis with nerve deafness. Arch Pathol 1966;82:506–517.

105 Churg J, Sherman RL: Pathologic characteristics of hereditary nephritis. Arch Pathol 1973;95:374–379.

106 Beathard GA, Granholm NA: Development of the characteristic ultrastructural lesion of hereditary nephritis during the course of the disease. Am J Med 1977;62:751–756.

107 Habib R, Gubler MC, Hinglais N, Noel LH, Droz D, Levy M, et al: Alport's syndrome: Experience at Hôpital Necker. Kidney Int Suppl 1982;11:S20–S28.

108 Hinglais N, Grünfeld JP, Bois E: Characteristic ultrastructural lesion of the glomerular basement membrane in progressive hereditary nephritis (Alport's syndrome). Lab Invest 1972;27:473–487.

109 Franco M, Schmitt F, Rejali WA, Viero RM, Bacchi CE: Renal interstitial foam cells are macrophages. Histopathology 1992;20:173–176.

110 Harris JP, Rakowski TA, Argy WPJ, Schreiner GE: Alport's syndrome representing as crescentic glomerulonephritis: A report of two siblings. Clin Nephrol 1978;10:245–249.

111   Rumpelt HJ, Steinke A, Thoenes W: Alport-type glomerulopathy: Evidence for diminished capillary loop size. Clin Nephrol 1992;37:57–64.

112   Heptinstall RH: Alport's syndrome, Nail-Patella syndrome, sickling disorders, and cyanotic congenital heart disease; in Heptinstall RH (ed): Pathology of the Kidney, ed 4. Boston, Little, Brown, 1992, pp 2045–2084.

113   Olsen JL, De Urdaneta AG, Heptinstall RH: Glomerular hyalinosis and its relation to hyperfiltration. Lab Invest 1985;52:387–398.

114   Rumpelt HJ: Hereditary nephropathy (Alport syndrome): Correlation of clinical data with glomerular basement membrane alterations. Clin Nephrol 1980;13:203–207.

115   Gaboardi F, Edefonti A, Imbasciati E, Tarantino A, Mihatsch MJ, Zollinger HU: Alport's syndrome (progressive hereditary nephritis). Clin Nephrol 1974;2:143–156.

116   Olson DL, Anand SK, landing BH, Heuser E, Grushkin CM, Lieberman E: Diagnosis of hereditary nephritis by failure of glomeruli to bind antiglomerular basement membrane antibodies. J Pediatr 1980;96:697–699.

117   Jenis EH, Valeski JE, Calcagno PL: Variability of anti-GBM binding in hereditary nephritis. Clin Nephrol 1981;15:111–114.

118   Savage CO, Pusey CD, Kershaw MJ, Cashman SJ, Harrison P, Hartley B, et al: The Goodpasture antigen in Alport's syndrome: Studies with a monoclonal antibody. Kidney Int 1986;30:107–112.

119   Melvin T, Kim Y, Michael AF: Amyloid P component is not present in the glomerular basement membrane in Alport-type hereditary nephritis. Am J Pathol 1986;125:460–464.

120   Yoshikawa N, Cameron AH, White RH: The glomerular basal lamina in hereditary nephritis. J Pathol 1981;135:199–209.

121   Farboody GH, Valenzuela R, McCormack LJ, Kallen R, Osborne DG: Chronic hereditary nephritis. A clinicopathologic study of 23 new kindreds and review of the literature. Hum Pathol 1979;10:655–668.

122   Antonovych TT, Deasy PF, Tina LU, D'Albora JB, Hollerman CE, Calcagno PL: Hereditary nephritis: Early clinical, functional, and morphological studies. Pediatr Res 1969;3:545–556.

123   Rumpelt HJ: Alport's syndrome: Specificity and pathogenesis of glomerular basement membrane alterations. Pediatr Nephrol 1987;1:422–427.

124   Yum M, Bergstein JM: Basement membrane nephropathy: A new classification for Alport's syndrome and asymptomatic hematuria based on ultrastructural findings. Hum Pathol 1983;14:996–1003.

125   Steffes MW, Barbosa J, Basgen JM, Sutherland DER, Najarian JS, Mauer SM: Quantitative glomerular morphology of the normal human kidney. Lab Invest 1983;49:82–96.

126   Kohaut EC, Singer DB, Nevels BK, Hill LL: The specificity of split renal membranes in hereditary nephritis. Arch Pathol Lab Med 1976;100:475–479.

127   Kimmelstiel P, Kim OJ, Beres J: Studies in renal biopsy specimens with the aid of the electron microscope. I. Glomeruli in diabetes. Am J Clin Pathol 1962;38:270–279.

128   Fujita S, Hayden RCJ: Alport's syndrome. Temporal bone report. Arch Otolaryngol 1969;90:453–466.

129   Arnold W: Überlegungen zur Pathogenese des cochleo-renalen Syndroms. Acta Otolaryngol (Stockh) 1980;89:330–341.

130   Johnsson LG, Arenberg IK: Cochlear abnormalities in Alport's syndrome. Arch Otolaryngol 1981;107:340–349.

131   Streeten BW, Robinson MR, Wallace R, Jones DB: Lens capsule abnormalities in Alport's syndrome. Arch Ophthalmol 1987;105:1693–1697.

132   Marks MI, Drummond KN: Benign familial hematuria. Pediatrics 1969;14:590–593.

133   Rogers PW, Kurtzman NA, Bunn SM, White MG: Familial benign essential hematuria. Arch Intern Med 1973;131:257–262.

134   Peterson AS, Schubert JJ: Benign hereditary nephritis. J Fam Pract 1977;4:437–441.

135   Eisenstein B, Stark H, Goodman RM: Benign familial haematuria in children from the Jewish communities of Israel: Clinical and genetic studies. J Med Genet 1979;16:369–372.

136   Gauthier B, Trachtman H, Frank R, et al: Familial thin basement membrane nephropathy in children with asymptomatic microhematuria. Nephron 1989;51:502–508.

137 Dische FE, Anderson VE, Keane SJ, Taube D, Bewick M, Parsons V: Incidence of thin membrane nephropathy: Morphometric investigation of a population sample. J Clin Pathol 1990;43:457–460.
138 Dische FE, Weston MJ, Parsons V: Abnormally thin glomerular basement membranes associated with hematuria, proteinuria or renal failure in adults. Am J Nephrol 1985;5:103–109.
139 Von Behrens Wieland: Mediterranean macrothrombocytopenia. Blood 1975;46:199–208.
140 Greaves M, Pickering C, Martin J, Cartwright I, Preston FE: A new familial 'giant platelet syndrome' with structural, metabolic, and functional abnormalities of platelets due to a primary megakaryocyte defect. Br J Haematol 1987;65:429–435.

Prof. Martin C. Gregory, Department of Medicine, University of Utah,
Salt Lake City, UT 84148 (USA)

Tryggvason K (ed): Molecular Pathology and Genetics of Alport Syndrome.
Contrib Nephrol. Basel, Karger, 1996, vol 117, pp 29–45

Application of Linked Markers for Genetic Diagnosis of Alport Syndrome

David F. Barker[a]*, Curtis L. Atkin*[b, c]*, Martin C. Gregory*[c, d]*,
Pamela R. Fain*[e]

Departments of [a]Physiology, [b]Biochemistry, [c]Internal Medicine,
[d]University Wasatch Clinics, [e]Medical Informatics, University of Utah
School of Medicine, Salt Lake City, Utah, USA

Most cases of Alport syndrome (AS) are caused by a defect of the COL4A5 gene, providing a direct approach to diagnosis with molecular genetic techniques. The diagnostic information may be used to detect affected individuals, avoiding risks and costs associated with renal biopsy. Genetic diagnosis may also be used for prenatal detection, for identification of asymptomatic carriers or testing of potential kidney donors who are themselves at risk. In addition, families with inherited nephritis caused by lesions in genes other than COL4A5 may now be reliably identified by excluding genetic linkage of their disease to markers around COL4A5.

Molecular genetic methods for the diagnosis of AS caused by defects in the COL4A5 gene include direct detection of mutations affecting the gene as well as the use of closely linked polymorphic markers for diagnosis based on genetic linkage. Mutation detection methods have the advantage of being applicable to individuals with no family history or to families with only a small number of samples available from affected and related family members. Diagnosis with genetic markers, in contrast, requires no detailed knowledge of the molecular defect, and can usually be achieved by gathering allele data at a limited number of highly informative linked or intragenic markers. The known spectrum of COL4A5 mutations causing AS suggests that gross, easily detectable changes such as deletions or other genomic rearrangements occur in about 10% of cases. The remainder appear to represent a substantial variety of point mutations affecting coding sequences, splicing signals and possibly other regulatory sequences. Routine diagnosis by direct mutation detection will require more rapid and routine methods for the detection of unknown point mutations within all of the coding and control elements needed for normal production of the COL4A5 protein. Adequate criteria for deciding whether any detected point change is deleterious must also be available. Until these advances are made, the genetic linkage approach will continue to be of diagnostic value. The current status of molecular methods for direct detection of COL4A5 mutations is discussed elsewhere in this volume. Applications for genetic linkage analysis using markers near COL4A5 are detailed here.

The Map of the X Chromosome near COL4A5

A reliable map of markers is essential for devising an efficient strategy for gathering critical genetic marker data in disease families and interpreting that data accurately. Knowledge of the order of markers with respect to each other and the disease gene and the genetic distances between the markers and the disease gene is needed. The map of the region of the X chromosome containing COL4A5 is well developed and based on information obtained by a variety of experimental approaches. Genetic mapping in large families with well-established AS has been useful for determining the genetic distance between individual markers and COL4A5 and often also provides some information about probe order. Genetic mapping data from linkage reference families, such as those distributed by the CEPH (Centre d'etude du polymorphisme humaine) can improve the precision of estimates for the genetic distance between markers and provide additional information about probe order.

The relative locations of markers and the nucleotide distances separating them may also be assessed by more direct physical methods. One of these is pulsed field electrophoresis, which permits the construction of maps of the positions of restriction sites which occur infrequently in human DNA. Recognition sites for restriction enzymes such as NotI and MluI occur at positions often separated by distances on the order of 1 million nucleotides (1 mb). Determining the position of marker probes with respect to such restriction sites, and orienting the map of sites with respect to the disease gene provides information on the order of the markers and their relative distances from the disease gene. Another useful physical mapping method takes advantage of naturally occurring or artificially induced chromosome rearrangements which involve unique breakpoints in the region of interest. By mapping the position of the probes and the disease gene with respect to these breakpoints, additional information is obtained regarding probe order and relative distance from the disease gene. A third method for establishing the physical proximity of markers involves the isolation of yeast artificial chromosome clones (YACs) which

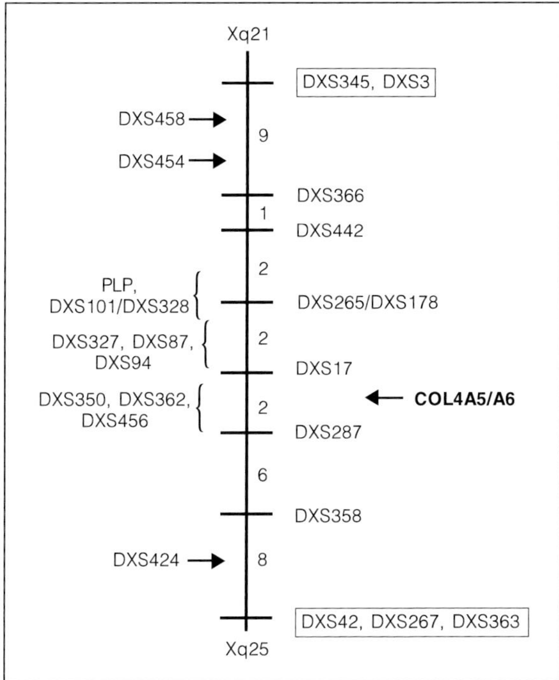

Fig. 1. Composite of genetic and physical map information for genetically useful markers in the vicinity of COL4A5/A6 derived as described in the text. Distances between ordered genetic markers are indicated in centimorgans (cM).

carry segments of human DNA as large as 1 mb. Isolation of a single YAC containing two or more markers establishes their physical proximity, when appropriate measures are taken to ensure that the YAC clone represents a single contiguous segment of the human genome. Subsequent refined mapping of the cloned segment can further establish the relative location of the markers. All of these methods have contributed to the summary map shown in figure 1.

Figure 1 is a composite of genetic and physical mapping information for the Xq21.3 to Xq25 region which surrounds the COL4A5 locus. The 'backbone' of figure 1 is a series of ordered genetic markers, with supporting relative odds of at least 100:1, and often greater than 1,000:1, from linkage studies in two large Alport families and the CEPH linkage reference families [1–3]. The genetic distances between the 'backbone' markers, as found in the combined data from the Alport and CEPH families, is indicated in cM. Two markers which are 2 cM apart, DXS17 and DXS287, are known to flank the position of the COL4A5 gene in the order shown, based on a chromosomal breakpoint map [1]. Markers indicated to the left in figure 1 are those which can be assigned an approximate location based on other available physical and/or genetic information. The markers DXS458 and DXS454 have been localized in the interval shown by analysis of recombinants found in XLA families [4] and CEPH families [3, 5]. The DXS328 locus is closely associated with DXS101, based on the finding that the DXS328 and DXS101 probes hybridize to the same 1,100 kb MluI, 250 kb BssHII and 270 kb EagI fragments [6]. DXS178 and DXS265 have been shown to lie within about 50 kb by identification of an SfiI fragment and several YAC clones that include both [6, 7]. The positions of DXS87, DXS94, DXS327 and DXS17 are distal to DXS178/265 by genetic mapping [2, 6], but their relative order is not well established. The map order of DXS350, DXS362, and the COL4A5 gene is also unknown; however, all lie in one interval defined by a set of specific chromosomal breakpoints [1].

The DXS287 marker is very close to and distal to COL4A5, based on chromosome breakpoint mapping [1] and supportive genetic evidence from CEPH families. The highly informative marker DXS456 has been localized to the same small interval as COL4A5, DXS350 and DXS362 by demonstrating PCR amplification of the human specific product from DNA of hybrid cell lines containing the same chromosomal segments used to map these markers [unpubl. data] (fig. 1). Evidence for two rare recombination events [unpubl. observations] indicates that DXS456 is telomeric to COL4A5, and is therefore a very tightly linked distal flanking marker. In general, the region distal to COL4A5 does not contain as high a density of useful markers as the proximal region; however, two markers of relatively high heterozygosity, DXS358 and DXS424, lie 6 and approximately 10 cM distal to COL4A5 respectively. A number of additional polymorphic markers have recently been identified and mapped to this region [3, 8]. Further refinements of their map localizations will likely provide additional useful markers for linkage studies.

Characteristics of Genetic Markers near COL4A5

Table 1 summarizes the properties of genetic markers shown in figure 1. Marker alleles may be detected by one of three methods. For loci defined by probes that detect a standard restriction fragment length polymorphism (RFLP), the probe name and the enzyme name(s) used to detect the RFLP(s) at that locus are given. Alleles at all RFLP loci may be determined by standard Southern blot hybridization. Purified genomic DNAs are cut with the appropriate restriction enzyme, separated on an agarose gel, transferred to a nylon filter and hybridized with a labeled probe. The probes mentioned in table 1

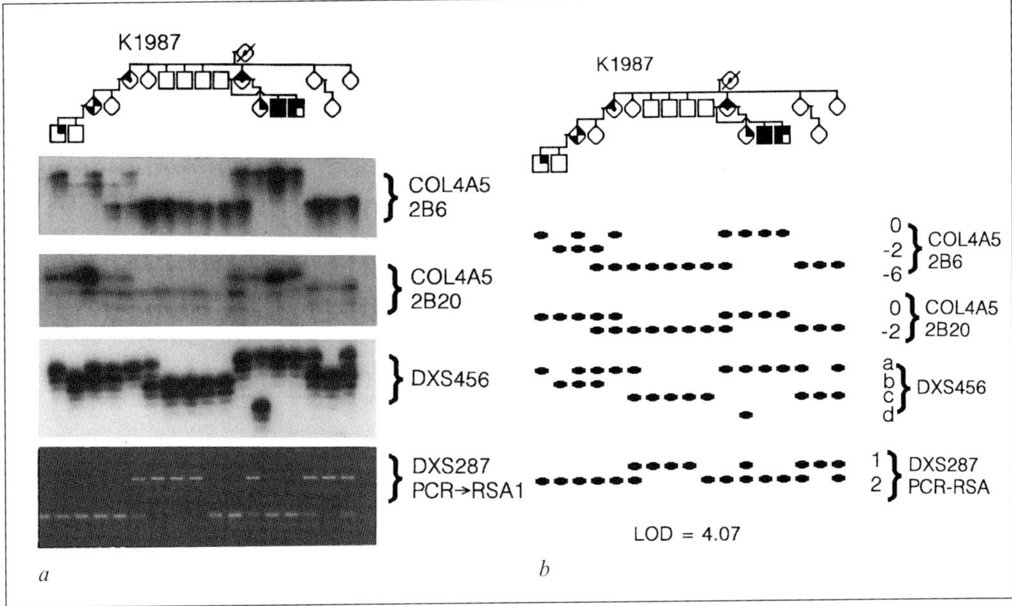

Fig. 2. Example of DNTR and PCR RFLP marker data in K1987. *a* Examples of data for markers 2B6, 2B20, DXS456 and DXS287 PCR + RSA. *b* Allele assignments are shown.

are available from the American Type Culture Collection (Rockville, MD, USA). For three of the RFLP loci, DXS17, DXS87 and DXS287, PCR primers that flank the position of the known polymorphic restriction site have been designed [16]. Primers flanking a polymorphic BsaHI (AhaII) site at the PLP gene locus have also been published [12]. The allelic forms at these sites may be determined by direct PCR amplification of DNA followed by digestion with the appropriate restriction enzyme and separation of the products on agarose gels. This is a simple and convenient assay method, as it does not require Southern blots and hybridization, the use of any radioactive components or DNA sequencing gels. An illustration of typing the DXS287 RsaI marker with this method is shown in figure 2.

A type of variation at the DNA level which has proven to be an excellent source of useful polymorphisms for genetic applications occurs at sequences which contain arrays of a tandemly reiterated simple sequence [18–20]. These arrays are virtually ubiquitous in the human and other mammalian genomes and are frequently found to have several allelic forms containing different

Locus	Probe/enzyme or marker type	Primers	HET %	Ref.
DXS3	DNTR	aatacataggtgtattgtgacc ccacctctctgaaagtgtgt	64	9
DXS3	p19–2/TaqI and MspI		47	1
DXS345	pRX72R1/TaqI or BglII		34	1
DXS458	DNTR	gataaaactgcatagaaatgcg caactgggatattgacattg	58	5, 10
DXS454	DNTR	agaagacataaggatactgc gatcccaactatttctttct	75	5, 10
DXS366	pRX329H2/TaqI		40	1
DXS442	pRX276E3/MspI		32	1
PLP	pJB010/MspI		46	11
PLP	pRL1/MspI		15	1
PLP	PCR followed by BsaHI (AhaII) digest	tctcgaattcccatgtcaatcatttt tctcgaattcgcacccgtaccctaactc	33	12
DXS328	pQST13R1/HindIII and PstI		45	1
DXS178	p212/9/TaqI		44	1
DXS178	DNTR	agttccaaacaaaatcccaaga attgtaaaaacttgatatttgct	65	13
DXS178	DNTR	tataatgagcatgcatca tgctcttcaggaaaattg	69	14
DXS178	DNTR	ccggaattctctacaataaatgacatt ccggtcgacgaatacaaacacagtatc	48	14
DXS178	DNTR	acagtacactgttctgca cctcaggaacctcttact	66	14
DXS265	pKZO33H4/MspI		45	1
DXS101	cX52.5/MspI		45	1
DXS101	TNTR	actctaaatcagtccaaatatct aaatcactccatggcacatgtat	80	15
DXS87	pG3–1/BglII		40	1
DXS87	PCR followed by BglII digest	aacttcacagagtttaggca cctacaaggctgtatgacca	Same RFLP	16
DXS94	pXG–12/PstI		44	1
DXS327	pQST7H1/MspI		48	1
DXS17	pS21/TaqI and MspI		35	1
DXS17	PCR followed by TaqI digest	gcaattatctgtattacttgaat ggtacatgacaatctcccaatat	Same RFLP	16
DXS350	pRX100M1/MspI		64	1
DXS362	pRX237E2/MspI		47	1

(continued next page)

Locus	Probe/enzyme or marker type	Primers	HET %	Ref.
COL4A5	2B-6 DNTR	tataatggaagttattcatgtagac gtgattcagatgttacttaaggac	76	17
COL4A5	2B-20 DNTR	ggtgccctttttaatacattttcc ttagacaagcttttactcccaaga	39	17
DXS456	DNTR	taactacacatgtgattctc taaagatagagtgactgatg	77	18
DXS287	pYNH3/RsaI		43	1
DXS287	PCR followed by RsaI digest	ggccgtcattagtgctgg gcaagatatgaatcctggag	Same RFLP	16
DXS358	pRX187M2/HindIII and TaqI		56	1
DXS424	DNTR	acctagttggaggctatgca cccagttactaacatctatg	83	5
DXS42	p43–15/BglII		24	1
DXS267	pchi16H2/MspI		55	1
DXS363	pRX258E1/TaqI		22	1

numbers of repeat elements, providing excellent informativeness for genetic analysis. The most common form is a dinucleotide tandem repeat (DNTR), usually of the repeating unit CpA. Allelic forms are typed by PCR assay, using primers flanking the repeat array. The highly informative markers DXS458, DXS454, DXS456 and DXS424 are all DNTRs. Typing of the DXS456 locus marker is also illustrated in figure 2. Several DNTRs have been developed for the DXS178 locus, in addition to the RFLP at this locus [13, 14]. A highly polymorphic trinucleotide repeat (TNTR) has been identified at the DXS101 locus [15]. This marker (table 1) is of particular value for linkage studies, because of its high heterozygosity, and also the generally better legibility of allele data obtained with PCR assay of arrays where the repeat size is greater than 2.

The DNTR data shown in figure 2 were obtained by labeling the PCR products with a single ^{32}P end-labeled primer in the reaction [18]. Internal labeling with ^{32}P nucleotides and unlabeled primers often produces autoradiographic patterns that are more difficult to interpret. Alternative methods [21] which detect only one strand of the synthesized product also give cleaner patterns. Resolution of the alleles, which usually differ in size by multiples of two basepairs, is accomplished on DNA sequencing gels or automated sequencing devices using fluorescent labels [20, 22].

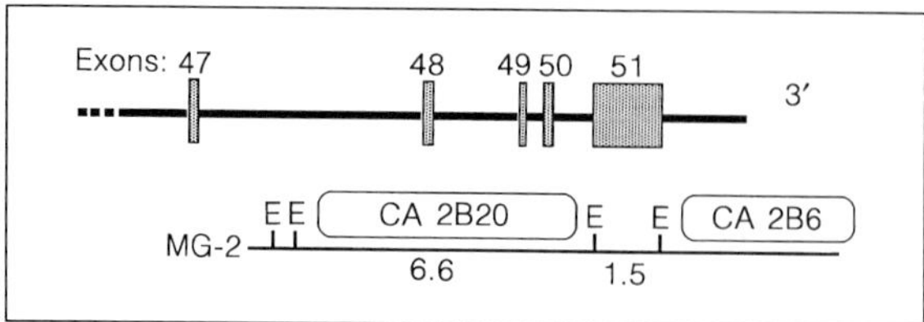

Fig. 3. Positions of the CA-containing segments corresponding to the DNTR markers 2B6 and 2B20 with respect to the coding segments of the COL4A5 gene.

DNTRs within COL4A5

Genetic polymorphisms within or immediately adjacent to the COL4A5 gene are the most reliable markers for this locus, prompting the search for simple sequence arrays in genomic segments which include COL4A5 exons. Two CpA length variants have been detected in a genomic clone spanning the 3' end of the COL4A5 gene [17]. Examples of allele data for these markers, designated 2B6 and 2B20, are shown in figure 2. The 2B6 and 2B20 markers are located at the positions indicated in figure 3 [unpubl. data]. The 2B6 marker is located within a few kb 3' of the last exon of COL4A5 and is heterozygous in about 76% of tested females. The second marker, 2B20, is located within a 6.6 kb EcoRI fragment that includes exons 48, 49 and 50. The heterozygosity of 2B20 is about 39%. The combined heterozygosity observed in CEPH mothers is about 79%, only slightly more than 2B6 alone. However, in studying Alport families, we have found specific instances where 2B20 is informative when 2B6 is not, suggesting the utility of testing both markers independently. With the recent isolation of YAC clones which contain the entire COL4A5 gene [23] it should soon be possible to identify additional simple sequence arrays in this region, providing additional useful polymorphisms for genetic studies.

Using Markers for Linkage Analysis

To obtain the most complete and reliable genetic data, the best strategy is simply to use the most highly informative markers that are closest to COL4A5 first, and then to apply secondary markers as needed for any carrier mothers who are not informative with the primary markers. We routinely apply the 2B6 and DXS456 markers first, which may provide full genetic informativeness. As secondary markers, we have used the intragenic 2B20 marker, a DXS178

DNTR or the DXS101 TNTR and the DXS287 RsaI and DXS17 TaqI markers in the PCR assay format. In diagnostically critical cases, it is important to identify informative markers flanking the COL4A5 locus in order to rule out the possibility of any rare, possibly intragenic, recombination event. Rare instances of DXS456 recombination with COL4A5 have been observed, although this marker is less than 1 cM from COL4A5. And, as indicated in figure 1, the DXS101 and DXS178 markers are 2–4 cM from the disease locus. Since marker testing also provides an opportunity to detect instances where inheritance patterns are inconsistent, the general confidence in linkage results with any particular family will increase with additional marker typing of that family.

Genetic Heterogeneity of AS

The strongest proof of COL4A5 linkage is detection of a gene defect resulting in a substantially altered protein. Detection of mosaicism for presence of the FNS1 antigen in the skin of a likely carrier female [24] may also be considered definitive. Such evidence is not routinely available for most Alport families and there is no well-established clinical distinction between patients with COL4A5 defects versus other gene defects which cause both autosomal recessive [25, 26] and autosomal dominant diseases with many similar phenotypes. Uncertainty exists regarding the relative prevalence of defects at the X-linked COL4A5 locus versus recessive defects involving COL4A3 and COL4A4 on chromosome 2 or other genes involved in autosomal forms of AS. This is partly due to the fact that COL4A5 mutations have as yet been found in only a fraction ($<25\%$) of families with diagnosed AS. With no available systematic, effective and routine methods for detection and recognition of all possible COL4A5 gene defects, the status of the majority of cases is not likely to be quickly resolved by direct mutation detection. In some families with autosomal forms of inherited nephritis, X-linkage may be ruled out by detecting instances of male-to-male transmission. However, many family histories will not include a sufficient sample to permit such an observation. In general, uncertainty about the probability that a particular case of AS may be due to an autosomal gene dictates caution in the application of molecular genetic methods for diagnostic purposes. In particular, diagnosis based on linked markers should only be undertaken after adequate proof of disease linkage to COL4A5 has been established. These precautions will be necessary until clinical methods exist for distinguishing the different forms of Alport and/or completely sensitive and calibrated molecular detection methods for mutations in COL4A5 are available.

Criteria for Establishing Linkage with Genetic Marker Data

Alport Kindred 1987 has been tested for several markers linked to COL4A5 and the data are shown in figure 2. The LOD score for linkage was calculated with the MLINK program of the LINKAGE software package [27] assuming parameter values described previously [1]. In K1987, inheritance of a haplotype of markers consisting of 2B6 allele '0', 2B20 allele '0', DXS456 allele 'a' and DXS287 allele '2' is completely associated with inheritance of the Alport phenotypes. The LOD score is 4.07, indicating greater than 10,000:1 odds for linkage of the disease to this region, assuming no a priori likelihood of such linkage. A LOD score of 3 is generally accepted as proof for establishing the localization for an inherited disease.

Figure 4 shows interpretations of marker data for a series of additional Alport kindreds. In each family shown in figure 4, inheritance of a specific marker haplotype is completely associated with the Alport phenotype. The calculated LOD scores vary from 1.16 to 3.85. For K2120, the observed LOD score of 3.85 establishes a very high degree of certainty for implicating the COL4A5 locus. However, even for kindreds with substantially lower LOD scores, the weight of genetic evidence is sufficient to conclude that linkage to the COL4A5 region is highly likely, because of the significant a priori probability that an inherited AS phenotype is due to a mutation in COL4A5, as summarized by table 2. Table 2 displays the posterior probabilities for COL4A5 nonlinkage, as a function of: (1) the established LOD score at a recombination fraction of zero and (2) the actual underlying (prior) probability that any family, similar in structure and clinical history to the one tested, carries a COL4A5 defect. The values are derived using a direct Bayesian analysis, based on the different possibilities for the prior probability and the potential LOD score observations. It is evident that LOD scores significantly less than 3 provide substantial confidence for concluding linkage. For example, if the assumption can be supported that the prior probability for linkage is greater than 0.75, then a LOD score of 1.0 provides about 97% confidence that the disease is associated with the COL4A5 locus.

The results shown in table 2 emphasize the importance of establishing the prior probability for linkage when interpreting the results of genetic linkage analysis in an individual family. This prior probability of linkage depends upon several factors. Accuracy of diagnosis is critical and reflects the effort that has been made to exclude other causes of nephritis and to confirm the diagnosis of AS by all available test methods [28]. The true frequency of unlinked families in the relevant population is also very important. Some population groups may have a higher frequency of the autosomal forms of AS. Additional studies are needed to estimate the frequency of unlinked fami-

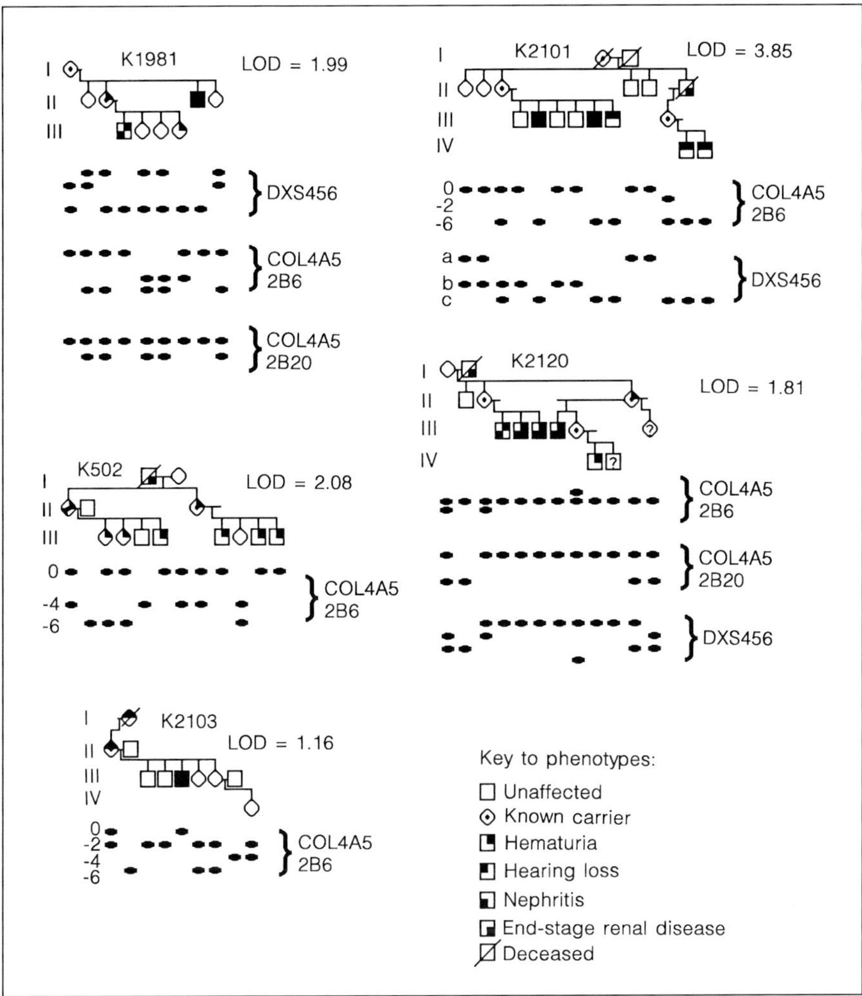

Fig. 4. Examples of LOD scores for families with marker patterns consistent with linkage of Alport disease phenotypes to the COL4A5 locus.

lies more precisely. Current estimates (see below) must be regarded as provisional. The higher the frequency of non-COL4A5 forms, the greater the chance that a particular family is unlinked and a correspondingly greater weight of consistent genetic evidence is needed to establish COL4A5 linkage. Families which appear to be 'typical' examples of X-linkage should be regarded with caution, as bias in ascertaining affected relatives is a strong possibility in cases of apparent AS. Obtaining as complete a family history as possible will improve

Table 2. Posterior probabilities that a genetically tested family has a disease gene that is unlinked to COL4A5, when the prior probability, 'Prior P(Link)', is as indicated and the listed LOD score (at theta = 0) is demonstrated[1]

	Prior P(Link)		0.95	0.90	0.75	0.50	0.25	0.10
LOD	Linkage Odds ↓	Prior → Odds	19:1	9:1	3:1	1:1	1:3	1:9
1.0	10:1		0.0052	0.010	0.032	0.090	0.23	0.47
1.5	31.6:1		0.0016	0.0035	0.010	0.031	0.087	0.22
2.0	100:1		0.0005	0.0011	0.0033	0.01	0.029	0.082
3.0	1,000:1		0.00005	0.00011	0.00033	0.001	0.003	0.0089

[1] Values were obtained by multiplying the associated odds ratios shown to obtain the combined odds ratio (L:N) for linkage vs. nonlinkage. The reported probability for nonlinkage is [N/(L + N)].

the opportunity for detecting instances of male-to-male transmission. The basis for diagnosis of such males should be carefully confirmed. For instance, alternate causes of single episodes of hematuria in any male offspring of affected males should be ruled out by repeated testing.

Detailed clinical studies of affected patients within unlinked families could eventually provide clinical criteria for distinguishing these cases from those with COL4A5 defects. In the meantime, accurate application of genetic linkage testing must depend on careful evaluation of the clinical characteristics and family history of each case. The value of maintaining a clinical database for supporting estimates of diagnostic probabilities is evident.

Possible Ambiguities in Linkage Data

As in all genetic diagnosis, ambiguities and contradictory patterns may emerge. The bulk of data may strongly favor linkage to COL4A5 in a particular family, but one or more apparent inconsistencies may be observed. Absence of a clinically detectable phenotype is common in female carriers, with approximately 7% of such carriers exhibiting no hematuria [1]. Other possible sources of confusion are sample-labeling errors or instances of nonpaternity. In both cases, the inconsistencies are due to incorrect assignments of familial relationships and testing of highly polymorphic autosomal markers will reveal and confirm the same inconsistencies.

In other cases, a new mutation in COL4A5 may be present in one of the individuals sampled. Close relatives of this individual may apparently carry the same chromosomal segment, inherited from the common ancestor, yet with no evidence of the disease. More confounding is the possibility that a new mutation may have arisen in the germ-line of a key individual, resulting in germ-line mosaicism. In such a case, two different offspring of the same parents, both inheriting identically marked chromosomal segments, may have a different disease status, strongly suggesting nonlinkage. Since new mutations in COL4A5 cause about 15% of all Alport-affected births [29] it is important to recognize when data supporting nonlinkage might also be interpreted as the occurrence of a new mutation, possibly involving germ-line mosaicism.

Autosomal Inheritance of AS

Autosomal recessive inheritance of AS has recently been conclusively demonstrated. The COL4A3/A4 loci on chromosome 2 encode two minor basement membrane collagen genes that are expressed in kidney, lung and muscle [30, 32], and the COL4A3 locus has previously been shown to encode the Goodpasture antigen [33]. Individuals who carry two mutated copies of COL4A3 or two mutated copies of COL4A4 appear to suffer from a form of AS that is indistinguishable from the X-linked form except that the severity of the disease in females is the same as in males [25, 26].

The existence of an autosomal dominant form of Alport has been controversial and most of the previously purported examples of such inheritance were subsequently shown to be instances of X-linkage. In view of the many possibly spurious causes for apparent inconsistencies with X-linkage that might suggest the existence of autosomal dominant AS, claims for such a finding must be carefully scrutinized. Figure 5 presents one kindred which is a very strong candidate for classification as autosomal dominant Alport. Figure 5a represents our current view of the family members and their clinical status. Figure 5b depicts the family as it was first studied. At that time it appeared to have a 'classic' X-linked appearance, with affected sons of two sisters having different degrees of affectedness. Initial genetic typing was done with RFLP probes for loci DXS336, DXS442, DXS94, DXS327, DXS350, DXS362 and DXS358. Distinct marker haplotypes were recognized and traced through the individuals as indicated. The inheritance patterns are consistent with the indicated familial relationships, however they are not consistent with linkage of the disease locus to these markers. Further investigation of the family history and specific clinical evaluation of at-risk individuals resulted in the elucidation of the extended structure shown in figure 5a. Three cases of

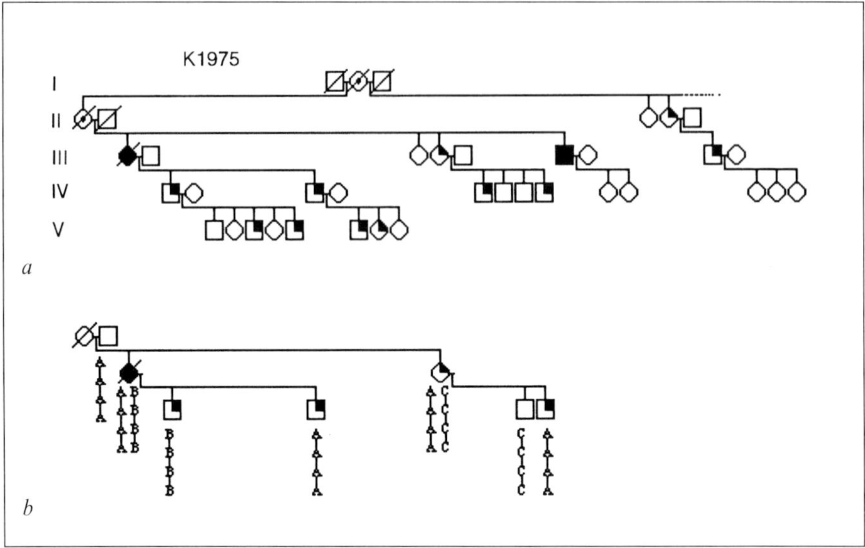

Fig. 5. Representation of K1975 illustrating possible dominant autosomal inheritance of an AS phenotype. *a* Our current view of the family structure, including three instances of male-to-male transmission. *b* The initial genetic linkage results obtained with a series of RFLP markers, as described in the text, which first showed the inconsistency with COL4A5 linkage of this disease phenotype. Symbols are as described in figure 4.

male-to-male transmission of the hematuria phenotype are evident, strongly suggestive of an autosomal dominant mode of inheritance. Alternatively, one might postulate an autosomal recessive mode of inheritance, with hematuria as a likely manifestation of the carrier state. Preliminary linkage data [unpublished observations in collaboration with S.T. Reeders] indicate no linkage to the COL4A3/A4 locus, however.

Estimating the Frequency of Hereditary Nephritis Which is Not Due to COL4A5

The definition of highly informative markers within and flanking the COL4A5 locus has considerably improved the power of genetic linkage studies to exclude the involvement of this locus. In the Utah study, there are currently over 70 families with evidence of hereditary glomerulonephritis, usually an apparently dominant form. Of these, 40 have a size and structure that should allow proof of genetic linkage to COL4A5, if it exists, as well as reliable exclusion of linkage. Exclusion of linkage might also be demonstrable with some of the smaller families, but they are not considered here. Of the 40 larger families, COL4A5 mutations are known for 8, genetic linkage to COL4A5 is established for a separate group of 9 and COL4A5 linkage appears to be excluded for 4. Nineteen are not completely evaluated. The tentative suggestion is that approximately 19% of AS families may carry lesions in gene(s) other than COL4A5. Since nearly all of the families examined have a structure consistent with X-linkage, some bias toward a low estimate must be considered. However, the incorrect exclusion of some families, because of data errors, complex phenomena or incomplete family history information, would tend to inflate the estimate of the non-COL4A5 fraction. Among the 40 families considered above, there are no strong candidates for recessive inheritance, although one study suggested that as much as 13% of AS might be due to recessive inheritance [25]. Further evaluation of the frequency of unlinked families in different populations and by different centers is needed. The frequency may vary in different ethnic groups or because clinical criteria vary systematically between different centers. Definition of the gene(s) involved in causing the disease in the families without any COL4A3, COL4A4 or COL4A5 defect would considerably aid classification efforts.

Conclusion

Techniques for complete evaluation of specific defects of the COL4A5 gene for diagnosing AS are advancing rapidly but are not yet perfected. The recent discovery of the COL4A6 gene which lies adjacent to COL4A5 and which might be mutated in some Alport families may further complicate direct mutation analysis at this locus. The method of genetic diagnosis with linked markers provides a suitable alternative in many cases and is readily performed and evaluated. Because of the high a priori likelihood of COL4A5 involvement in causing AS, a modest amount of consistent genetic data provides high confidence regarding COL4A5 linkage. Families and individuals with a phenotype that strongly resembles AS, but without any genetic lesion in COL4A5, are known to exist and may account for about 19% of inherited nephritis. An unknown fraction of these may be instances of inheritance of COL4A3 or COL4A4 defects. Appropriate precautions must be taken to ensure that such families are not improperly diagnosed with COL4A5 linked markers. Finding the genetic locations and all of the genes involved in causing these related syndromes would further improve our diagnostic capabilities.

References

1   Barker DF, Fain PR, Goldgar DE, Dietz-Band JN, Turco AE, Kashtan CE, Gregory MC, Tryggva-
     son K, Skolnick Atkin CL: High density genetic and physical mapping of DNA markers near the
     X-linked Alport syndrome locus: Definition and use of flanking polymorphic markers. Hum Genet
     1991;88:189–194.
2   Fain PR, Luty JA, Guo Z, Nguyen K, Barker DF, Litt M: Localization of the highly polymorphic
     microsatellite DXS456 on the genetic linkage map of the human X chromosome. Genomics
     1991;11:1155–1157.
3   Fain PR, Kort EN, Chance PF, Nguyen K, Redd DF, Econs MJ, Barker DF: A 2D crossover-
     based map of the human X chromosome as a model for map integration. Nat Genet 1995;9:261–265.
4   Parolini O, Conley ME: Additional polymorphisms useful in linkage analysis for X-linked agamma-
     globulinemia. Immunodeficiency 1993;4:217–219.
5   Huang TH-M, Cottingham RW, Ledbetter DH, Zoghbi HY: Genetic mapping of four dinucleotide
     repeat loci, DXS453, DXS458, DXS454 and DXS424, on the X chromosome using multiplex
     polymerase chain reaction. Genomics 1992;13:375–380.
6   Parolini O, Hejtmancik JF, Allen RC, Belmont JW, Lassiter GL, Henry MJ, Barker DF, Conley
     ME: Linkage analysis and physical mapping near the gene for X-linked agammaglobulinemia at
     Xq22. Genomics 1993;15:342–349.
7   Willard HF, Cremers F, Mandel JL, Monaco AP, Nelson DL, Schlessinger D: Report and Abstracts
     of the Fifth International Workshop on Human X Chromosome Mapping. Cytogenet Cell Genet
     1994;67:295–358.
8   Gyapay G, Morissette J, Vignal A, Dib C, Fizames C, Millasseau P, Marc S, Bernardi G, Lathrop
     M, Weissenbach J: The 1993–1994 Genethon human genetic linkage map. Nat Genet 1994;7:246–
     339.
9   Stanier P, Newton R, Forbes SA, Ivens A, Moore GE: Polymorphic dinucleotide repeat at the
     DXS3 locus. Nucleic Acids Res 1991;19:4793.
10   Weber JL, Kwitek AE, May PE, Polymeropoulos MH, Ledbetter S: Dinucleotide repeat polymorph-
     isms at the DXS453, DXS454 and DXS458 loci. Nucleic Acids Res 1990;18:4037.
11   Raskind WH, Wolff J, Hudson LD, Bird TD: RFLP detected by a genomic probe from the human
     X-linked proteolipid protein gene, PLP. Hum Mol Genet 1992;1:288.
12   Trofatter JA, Pratt VM, Dlouhy SR, Hodes ME: AhaII polymorphism in human X-linked proteolipid
     protein gene. Nucleic Acids Res 1991;19:6057.
13   Allen RC, Belmont JW: Dinucleotide repeat polymorphism at the DXS178 locus. Hum Mol Genet
     1992;1:216.
14   de Weers M, Mensink RGJ, Kenter M, Schuurman RKB: Three dinucleotide repeat polymorphisms
     at the DXS178 locus. Hum Mol Genet 1992;1:653.
15   Allen RC, Belmont JW: Trinucleotide repeat polymorphism at DXS101. Hum Mol Genet 1993;2:
     1508.
16   Kornreich R, Astrin KH, Desnick RJ: Amplification of human polymorphic sites in the X-chromo-
     some region q21.33 to q24: DXS17, DXS87, DXS287 and alpha-galactosidase A. Genomics 1992;
     13:70–74.
17   Barker DF, Cleverly J, Fain PR: Two CA-dinucleotide polymorphisms at the COL4A5 (Alport
     syndrome) gene in Xq22. Nucleic Acids Res 1992;20:929.
18   Luty JA, Guo Z, Willard HF, Ledbetter DH, Ledbetter S, Litt M: Five polymorphic microsatellite
     VNTRs on the human X chromosome. Am J Hum Genet 1990;46:776–783.
19   Weber JL, May PE: Abundant class of human DNA polymorphisms which can be typed using the
     polymerase chain reaction. AM J Hum Genet 1989;44:388–396.
20   Edwards A, Civitello A, Hammond HA, Caskey CT: DNA typing and genetic mapping with trimeric
     and tetrameric tandem repeats. Am J Hum Genet 1991;49:746–756.
21   Browne D, Barker D, Litt M: Dinucleotide repeat polymorphisms at the DXS365, DXS443 and
     DXS451 loci. Hum Mol Genet 1992;1:213.
22   Litt M, Hauge X, Sharma V: Shadow bands seen when typing polymorphic dinucleotide repeats:
     Some causes and cures. Biotechniques 1993;15:280–284.

23 Vetrie D, Flinter F, Bobrow M, Harris A: Construction of a yeast artificial chromosome contig encompassing the human alpha5(IV) collagen gene (COL4A5). Genomics 1992;14:634–642.

24 Kleppel MM, Kashtan C, Santi PA, Wieslander J, Michael AF: Distribution of familial nephritis antigen in normal tissue and renal basement membranes of patients with homozygous and heterozygous Alport familial nephritis. Lab Invest 1989;61:278–289.

25 Lemmink HH, Mochizuki T, van den Heuvel LPWJ, Schröder CH, Barrientos A, Monnens LAH, van Oost BA, Brunner HG, Reeders ST, Smeets HJM: Mutations in the type IV collagen α3 (COL4A3) gene in autosomal recessive Alport syndrome. Hum Mol Genet 1994;3:1269–1273.

26 Mochizuki T, Lemmink HH, Mariyama M, Antignac C, Gubler MC, Pirson Y, Verellen-Dumoulin C, Chan B, Schröder CH, Smeets HJM, Reeders ST: Identification of mutations in the α3 and α4 type IV collagen genes in autosomal recessive Alport syndrome. Nat Genet 1994;8:77–81.

27 Lathrop GM, Lalouel JM, Julier C, Ott J: Multilocus linkage analysis in humans: Detection of linkage and estimation of recombination. Am J Hum Genet 1985;37:482–498.

28 Reeders ST: Molecular genetics of hereditary nephritis. Kidney Int 1992;42:783–792.

29 Gregory MC, Atkin CL: Alport syndrome; in Schrier RW, Gottschalk CW (eds): Diseases of the Kidney. Boston, Little, Brown, 1993, pp 571–591.

30 Mariyama M, Leinonen A, Mochizuki T, Tryggvason K, Reeders ST: Complete primary structure of the human α3(IV) collagen chain: Coexpression of the α3(IV) and α4(IV) chains in human tissues. J Biol Chem 1994;269:23013–23017.

31 Leinonen A, Mariyama M, Mochizuki T, Tryggvason K, Reeders ST: Complete primary structure of the human type IV collagen alpha-4(IV) chain: Comparison with structure and expression of the other alpha(IV) chains. J Biol Chem 1994;269:26172–26177.

32 Mariyama M, Zheng K, Yang–Feng TL, Reeders ST: Colocalization of the genes for α3(IV) and α4(IV) chains of type IV collagen to chromosome 2 bands q35–q37. Genomics 1992;13:809–813.

33 Turner N, Mason PJ, Brown R, Fox M, Povey S, Rees A, Pusey CD: Molecular cloning of the human Goodpasture antigen demonstrates it to be the alpha-3 chain of type IV collagen. J Clin Invest 1992;89:592–601.

David F. Barker, PhD, Department of Physiology, School of Medicine, The University of Utah, 410 Chipeta Way, Room 156, Research Park, Salt Lake City, UT 84108 (USA)

Tryggvason K (ed): Molecular Pathology and Genetics of Alport Syndrome.
Contrib Nephrol. Basel, Karger, 1996, vol 117, pp 46–79

Molecular Properties of the Glomerular Basement Membrane

Taina Pihlajaniemi

Collagen Research Unit, Biocenter, and Department of Medical Biochemistry,
University of Oulu, Finland

The evolution of extracellular matrices was a prerequisite for the development of multicellular organisms, and a special entity among extracellular matrices consists of basement membranes, thin sheet-like structures ubiquitously distributed in the body. In most tissue locations they separate epithelial and endothelial cells from the underlying stroma, but in certain locations such as the glomeruli of the kidney they separate opposing cell layers. They also surround individual adipocytes, muscle cells and nerve cells. Basement membranes often form continuous sheets, but may be discontinuous in some locations. Their thickness varies between 20 and 350 nm, being usually 60–80 nm. Basement membranes provide physical support for tissues and serve to compartmentalize them. They also function in many critical biological processes, including passive molecular sieving, cell attachment and migration, development, tissue regeneration and repair, maintenance of cell polarization, and as a reservoir of growth factors, enzymes and plasma proteins.

The molecular and cellular aspects of basement membranes under normal and pathological conditions have been extensively reviewed by Timpl and Dziadek [1], and more recently in a book by Rohrbach and Timpl [2, see reviews therein] and are reviewed further in this volume. Numerous other reviews also exist on the structure, function and pathology of basement membranes [3–5] or their major components, type IV collagen [6, 7], laminin [8–11] and nidogen/entactin [12]. In general, basement membranes are important in providing physical support for tissues and a substratum for cell attachment. The renal glomerular basement membrane (GBM) has an additional important role in the glomerular filtration of macro-

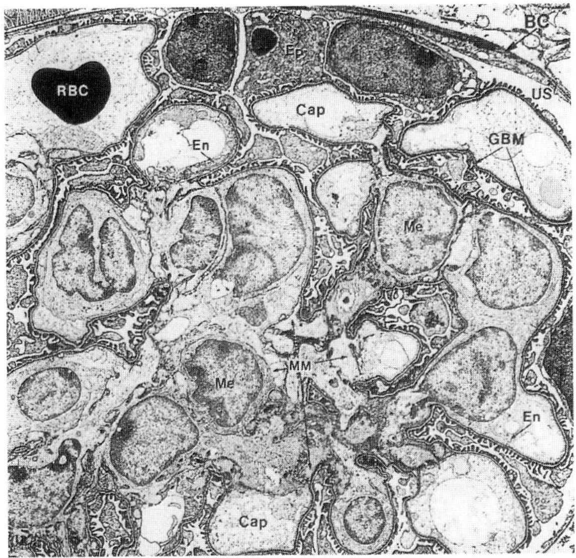

Fig. 1. Electron micrograph of a section of the kidney glomerulus. The glomerulus is composed of three cell types, the endothelial (En), epithelial (Ep) and mesangial (Me) cells, the glomerular basement membrane (GBM) and the mesangial matrix (MM) surrounded by the Bowman's capsule (BC). A red blood corpuscle (RBC) is seen in one of the capillaries. Urinary space (US), capillary (Cap). × 1,750. [Reproduced from 14 with permission.]

molecules. This review will focus on the molecular structure of the GBM, but data on basement membranes from other sources are by necessity also included.

Structure and Function of the Renal Glomerulus

The human kidney contains about 0.8–1.2 million nephrons, each composed of a renal corpuscle, the glomerulus surrounded by Bowman's capsule, and a tubular portion [13]. The first step in the formation of urine is the ultrafiltration of plasma in the renal glomerulus. The blood enters the glomerulus through its afferent arteriole, which branches into specialized capillary loops, the glomerular tuft, that are in association with mesangial cells and a mesangial matrix (fig. 1). Capillary filtration takes place into the space between the capillaries and Bowman's capsule, the latter consisting of an inner epithelial cell layer surrounded by a basement membrane [14]. The filtration unit com-

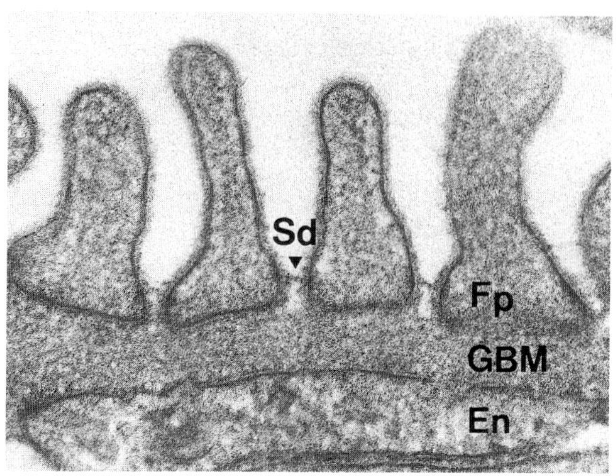

Fig. 2. Electron micrograph of the capillary filtration unit. The glomerular basement membrane (GBM) is seen sandwiched between the fenestrated capillary endothelium (En) and the epithelial foot processes (Fp) with intervening slit diaphragm (Sd). × 103,200. [Electron micrograph courtesy of Dr. Sadayuki Inoue, Department of Anatomy and Cell Biology, McGill University, Montreal, Canada.]

prises the GBM sandwiched between endothelial and epithelial cells. The endothelial cells of the glomerular capillaries contain fenestrae 40–70 nm in diameter that permit direct contact between the blood constituents and the GBM [15]. The epithelial cells, also called podocytes, are attached to the GBM by interdigitating foot processes, which are connected to one another by slit diaphragms between 25 and 60 nm in diameter, resulting in a zipper-like appearance [16]. The unusual thickness of the GBM, about 350 nm, is thought to be due to fusion of the endothelial and epithelial basement membranes during development [3, 14]. Electron microscopy of most basement membranes usually reveals a two-layered structure with a translucent layer, the lamina rara, in contact with the cells, and a more electron-dense layer, the lamina densa, towards the underlying matrix. The lamina rara may be lacking, as in the cornea. In many examinations the GBM appears to have three layers, a subendothelial electron-translucent layer, termed the lamina rara interna, an electron-dense central layer, the lamina densa, and a subepithelial electron-translucent layer, the lamina rara externa [14]. This triple structure may be an artifact of the processing of samples for electron microscopy, however, [17], as a uniform appearance can also be found (fig. 2).

During the filtration process the plasma passes successively through the endothelial fenestrae, the GBM and the epithelial slit diaphragms, after which it enters the urinary space [14]. Under physiological conditions, cellular elements and most plasma proteins are retained in the capillary, while water, nitrogenous waste products, electrolytes and low-molecular-weight proteins enter the urinary space. The porosity and charge of the GBM has been shown to be critical for the filtration process, and it regulates filtration in a size-selective fashion as follows: filtration of 32 kD > 62 kD > > 125 kD [18]. Furthermore, it acts as a charge-selective barrier so that negatively charged proteins are filtered less readily than neutral ones and cationic proteins have the easiest passage [19].

Many acquired and genetic diseases are known to affect the GBM, resulting in loss of serum proteins into the urine and impairment of the charge-selective filtration process [1, 3, 5–7, 20, 21]. Basement membrane abnormalities are thought to play a major role in the development of the secondary manifestations in diabetes, and Goodpasture's syndrome represents an autoimmune disease in which antibodies to intrinsic basement membrane components are illicited, the immunological reactions leading to glomerulonephritis and pulmonary hemorrhage. Other immunologically mediated glomerulonephritides are also known. GBM abnormalities also characterize the various forms of Alport's syndrome, polycystic kidney disease and the congenital nephrotic syndromes.

Glomerular Basement Membrane Components

The molecular composition of the GBM is in essence similar to that of basement membranes in other tissues, but it also posessess certain specific features. Like other basement membranes, it contains type IV collagen, laminin, proteoglycans and nidogen/entactin as its major components, although many more components have been estimated to occur in lesser quantities [1, 4]. Tissue-specific isoforms and differences in the ratios of the major components are now recognized, particularly for type IV collagen and laminin, and basement membranes with unique compositional and functional characteristics therefore exist.

Type IV Collagen
The major structural component of the GBM is type IV collagen, a protein belonging to the large family of collagens, at present known to contain nineteen distinct types [22, 23]. Each type IV collagen molecule consists of three polypeptide chains, termed α chains. Six genetically distinct type IV collagen α chains, $\alpha 1(IV)$–$\alpha 6(IV)$, have been identified, which signifies that more than one form of assembly of type IV molecules must exist in tissues [22–24].

Furthermore, the α3(IV) chain has been found to be subject to alternative splicing which generates variant polypeptides [25]. The most abundant chain combination, and the one present in most basement membranes, involves a molecule with two α1(IV) chains and one α2(IV) chain, [α1(IV)]₂α2(IV), but homotrimers of α1(IV) chains have also been found [1, 4, 7]. Other subclasses of type IV collagen must exist, although it is not fully known in what combinations the α3(IV), α4(IV), α5(IV) and α6(IV) chains occur. A largely separate network has been proposed for the novel chains and at least [α3(IV)]₂α4(IV) heterotrimers and [α3(IV)]₃ homotrimers appear to exist in the GBM [26–28].

The tissue distribution of type IV collagen is restricted to basement membranes. The stoichiometry of the α1(IV)–α5(IV) chains differs among tissues, the α1(IV) and α2(IV) chains occurring in all basement membranes while the others appear to have a more restricted distribution [see 29 and references therein] (table 1). All five chains are found in the kidney. The α1(IV) and α2(IV) chains appear to be restricted to the subendothelial regions in the adult GBM, in addition to being present in the mesangial matrix, Bowman's capsule and tubular and vascular basement membranes. The α3(IV) and α4(IV) chains appear uniformly across the GBM and in Bowman's capsule and the distal tubular basement membrane, while the α5(IV) chain is thought to be restricted to the GBM in the kidney, and the location of the most recently discovered component, the α6(IV) chain, awaits elucidation.

Structurally, α chains can be divided into three domains: 7S, a minor collagenous domain in the NH₂ terminus rich in cysteines and lysines; a major collagenous domain; and NC1, a noncollagenous segment present at the COOH end of each polypeptide [1, 4]. The collagenous domains comprise a thread-like segment about 400 nm long with the NC1 domain forming a large globule at the COOH terminus. The collagenous sequences consist of repeating Gly-Xaa-Yaa amino acid triplets interrupted at several sites by short noncollagenous sequences. The presence of the smallest amino acid, glycine, as every third residue is essential to allow the three α chains to come together and form the characteristic collagen triple helix. The complete primary structure of the α1(IV) chain has been determined for man [30, 31], the mouse [32], *Drosophila* [33], the sea urchin [34] and *Caenorhabditis elegans* [35], that of the α2(IV) chain for man [36], the mouse [37], *Ascaris suum* [38] and *C. elegans* [39] and that of the α3(IV) chain [40], the α4(IV) chain [41], the α5(IV) chain [42, 43] and the α6(IV) chain [44, 45] for man (table 1). The human α1(IV) chain consists of a 27-residue signal peptide, a 15-residue NH₂-terminal noncollagenous sequence, a 1398-residue collagenous sequence with 21 interruptions of 2–11 residues, and a 229-residue NC1 domain. The other type IV collagen chains show similar arrangements, and depending on their extent of homology the chains can be divided into an α1(IV)-like class containing the α1(IV), α3(IV) and α5(IV) chains and an α2(IV)-like class containing the α2(IV), α4(IV) and α6(IV) chains. The locations of potential site for asparagine-linked glycosylation in the 7S domain and cysteine residues are conserved in the various chains, although the exact number of the latter residues varies slightly. The overall sequence homology with respect to the collagenous region of the various chains is not high, but the positions of most of the interruptions, including all the large ones, are conserved between the chains. Thus, the lengths of the Gly-Xaa-Yaa domains are well conserved, although some length variation occurs in the repeating Gly-Xaa-Yaa sequences separating the interruptions. The α1(IV)–α6(IV) chains show high homology with respect to their NC1 domains, indicating that the function of these domains necessitates sequence conservation. These domains vary between 227 and 231 residues, and they all are characterized by two homologous symmetrical halves and 12 completely conserved cysteine residues which participate in the formation of intramolecular and intermolecular disulfide bonds [for details, see 46].

The synthesis of collagen is a complex process involving many cotranslational and posttranslational modifications catalyzed by at least nine specific enzymes and several nonspecific ones [47]. The processing can be regarded as taking place in two stages. The intracellular modifications include: (1) removal of the signal peptide; (2) hydroxylation of many of the Yaa-position proline and lysine residues to 4-hydroxyproline and hydroxylysine; (3) hydroxylation of a few Xaa-position prolines to 3-hydroxyproline; (4) addition of galactose, or galactose and then glucose, to some of the hydroxylysine residues; (5) glycosylation of certain asparagines; (6) association of the COOH-terminal noncollagenous domains through a process regulated by the structure of these domains, and (7) formation of both intrachain and interchain disulfide bonds. Following secretion into the extracellular space, the NH_2 and COOH-terminal propeptides may be cleaved by N-proteinase and C-proteinase, respectively. Next, the collagen molecules self-assemble into fibrils or other supramolecular aggregates. Finally, lysyl oxidase catalyzes the formation of lysine and hydroxylysine-derived cross-links between adjacent molecules. Synthesis of type IV collagen involves most of the same cotranslational and posttranslational modifications as outlined above. Intracellular chain selection and the alignment of the correct three polypeptide chains is thought to be mediated by the NC1 domains, and folding of the triple helix proceeds from the COOH-terminal end. Intrachain disulfide bond formation within the NC1 domains is probably required, but intramolecular interchain links are not formed [1, 4]. Type IV collagen molecules are characterized by numerous hydroxylysine residues, and the presence of 3-hydroxyproline which has not been found in any other mammalian proteins [47]. In contrast to fibril-forming collagens, type IV

Table 1. Structure and tissue distribution of basement membrane components

Poly-peptide chain	Full-length cDNA-derived sequences		Molecular weight kD	Tissue distribution	Chain combinations
	species	length in amino acids			
Type IV collagen					
$\alpha1(IV)$	Human	1642	158	Almost all basement membranes	$\alpha1(IV)_2\alpha2$, $\alpha1(IV)_3$, and probably $\alpha3(IV)_2\alpha4$ and $\alpha3(IV)_3$
	Mouse	1642			
	Drosophila	1752			
	Sea urchin	1630			
	C. elegans	1731			
$\alpha2(IV)$	Human	1676	164	Almost all basement membranes	
	Mouse	1679			
	Ascaris suum	1737			
	C. elegans	1733			
$\alpha3(IV)$	Human	1642	162	Basement membranes of brain, cochlea, eye, GMB and Bowman's capsule in kidney, lung and synapses	
$\alpha4(IV)$	Human	1652	164	As above	
$\alpha5(IV)$	Human	1659	153	Many tissues	
$\alpha6(IV)$	Human	1657	161	At least choroid plexus, esophagus, heart, kidney, meninges, placenta, skeletal muscle	
Laminin					
$\alpha1$	Human	3058	337	Restricted expression, mainly in kidney and arteries	$\alpha1\beta1\gamma1$ (EHS-laminin or laminin 1) $\alpha2\beta1\gamma1$ (merosin or laminin 2)
	Mouse	3060			
$\alpha2$	Human	3088	343	Many basement membranes; not in kidney	$\alpha1\beta1\gamma1$ (s-laminin or laminin 3) $\alpha2\beta2\gamma1$ (merosin/s-laminin or laminin 4)

α3	Human	1693	165	Epithelial basement membranes	α3β3γ2 (kalinin/epiligrin or laminin 5)
β1	Human	1765	198	Almost all basement membranes	α3β1γ1 (K-laminin or laminin 6)[a]
	Mouse	1768			
	Drosophila	1758			
β2	Human	1766		Many basement membranes; strongly expressed in synaptic basement membranes of motor neurons, and perineural, glomerular and arterial basement membranes	
	Rat	1766	190		
β3	Human	1153	127	Epithelial basement membranes	
β4				Identified in a chicken eye cDNA library	
γ1	Human	1576	178	Almost all basement membranes; most abundant laminin chain	
	Mouse	1574			
	Drosophila	1606			
γ2	Human	1172/1090[b]	129/119	Epithelial basement membranes	
	Mouse	1174			
Nidogen/ entactin	Human	1219	141	All basement membranes	
	Mouse	1217			
Perlecan	Human	4372	467	All basement membranes	
	Mouse	3707			

Amino acid numbers are given only if full-length sequences are known. Partial sequences, which may be known for some of the components (see text), are not shown. The polypeptide length is given for the mature protein, and the NH_2-terminal end has either been determined by protein sequencing or by assignment of the most probable signal peptide cleavage site. The molecular weights are cDNA-derived if full-length sequences are available and thus lack the effect that various posttranslational modifications, such as glycosylation, may have on the molecular weight of the corresponding native protein.

[a] Not fully determined that α3 in K-laminin is identical to α3 in kalinin.

[b] Two variant structures encoded by cDNA clones.

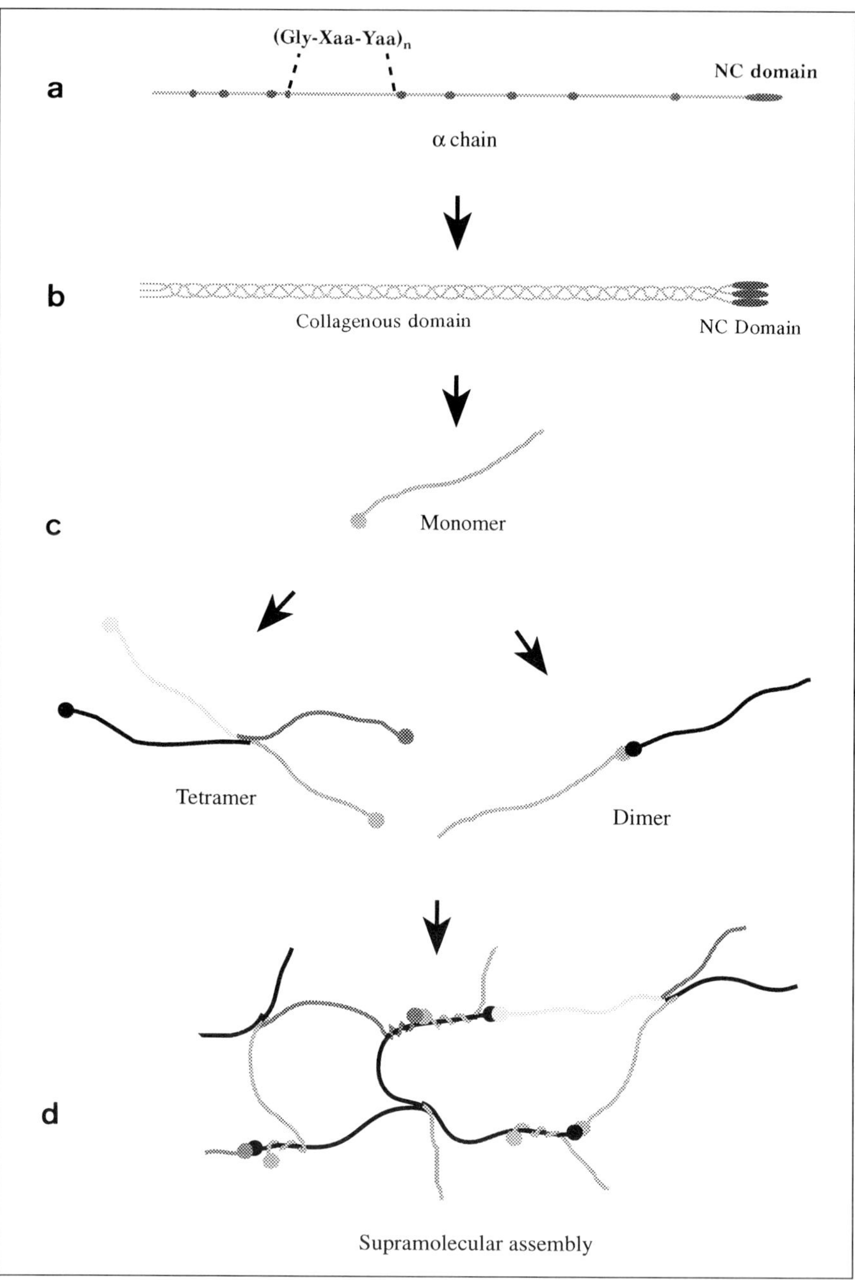

Fig. 3. Schematic representation of the assembly of the type IV collagen molecules. *a* A type IV collagen α chain containing short interruptions (●) in the repeating Gly-Xaa-Yaa sequence. *b* A type IV collagen molecule (monomer, shown in black and different shades

collagen molecules are not processed proteolytically following secretion from the cell [1, 4].

The secreted type IV collagen molecules self-assemble into a complex network [48, 49]. This collagen is covalently cross-linked by disulfide bonds and nonreducible lysyl oxidase-derived bonds and is thus not readily extractable, but in vitro self-assembly experiments and in situ observation of network structures have provided information about the nature of the aggregates. Type IV collagen molecules (monomers) form linear dimers through association between the NC1 domains of two molecules (fig. 3). This association is stabilized by disulfide bond formation, resulting in a stable hexameric NC1 complex. It is not clear whether this event is intracellular or extracellular. Tetrameric structures are formed by alternate parallel and antiparallel association of the NH_2-terminal ends of four triple helical monomers. This structure forms initially through noncovalent interactions and is subsequently stabilized by disulfide bridges and lysine-derived nonreducible cross-links. The four-armed tetramer is resistant to proteolytic degradation and was termed the 7S domain due to its sedimentation coefficient. This structure is also characterized by carbohydrates, which probably participate in restricting the interactions between the subunits. The bovine GBM 7S domain has been found to contain 12 complex oligosaccharides (one in each chain) at positions 126 and 138 for the α1(IV) and α2(IV) chains, respectively [50]. The dimers and tetramers then assemble further to form a network of highly branched filaments that include laterally aligned molecules along the length of the triple helical domains and molecules twisting around each other [49]. The result is a three-dimensional tight-meshed network that serves as the major structural support for the basement membrane [1, 4, 51]. Other basement membrane components bind to this collagen network (see below).

Electron microscopy of type IV collagen consisting of α1(IV) and α2(IV) chains, coupled with amino acid sequence data, has shown that the interruptions in the repeated Gly-Xaa-Yaa sequence coincide with kinks in the otherwise rigid triple helical segments [1, 4]. Thus the interruptions modulate the

of gray to visualize more clearly separate molecules) is composed of three α chains forming a triple helical structure. Since six type IV collagen α chains exist, monomers of different chain compositions exist (see text). *c* Monomers associate through their COOH-terminal noncollagenous domains forming dimers, and through their NH_2-terminal ends forming tetramers. *d* A complex network is formed through COOH-terminal dimerization, NH_2-terminal tetramer formation and lateral association and intertwining of triple helical domains. [Figure courtesy of Dr. Karl Tryggvason, Department of Biochemistry, University of Oulu, Finland.]

flexibility of the molecules and are essential to allow the formation of a network rather than more rigid structures such as fibrils. The interruptions may serve other functions as well, such as acting as binding sites for other basement membrane components.

In addition to providing mechanical strength for the basement membranes and serving as a scaffold for the binding of other components, type IV collagen is involved in cell binding. One major cell binding site has been located about 100 nm away from the NH_2 terminus of the molecule, the cellular receptors interacting with this region being integrins $\alpha 1\beta 1$ and $\alpha 2\beta 1$ [52, 53]. Integrins are a family of transmembrane receptors consisting of α and β subunits which play a major role in the adhesion of cells to extracellular matrix components [54]. The $\alpha 1\beta 1$ recognition site on $\alpha 1(IV)_2\alpha 2(IV)$ collagen has been pinpointed to an $\alpha 1(IV)$ chain aspartate and an $\alpha 2(IV)$ chain arginine, each located at about amino acid position 460 and embedded in a short triple helical segment stabilized by disulfide bonds [55].

Mutations in the genes for the $\alpha 1(IV)$ and $\alpha 2(IV)$ chains are only known for *C. elegans*. An embryonic lethal phenotype, *emb*-9, is seen in animals with mutations in the $\alpha 1(IV)$ gene [35], and embryonic lethality, *let*-2, is likewise observed in the $\alpha 2(IV)$ gene [39]. In view of the ubiquitous occurrence of the $\alpha 1(IV)$ and $\alpha 2(IV)$ chains in basement membranes, it may be possible that mutations for these chains cannot be identified in higher organisms due to their early lethal effect. On the other hand, mutations in the type IV collagen chains with more restricted occurrence are associated with certain human diseases. Preliminary evidence suggests a linkage between the rare autosomal forms of Alport's syndrome and the $\alpha 3(IV)$ and $\alpha 4(IV)$ genes, located in close association on chromosome 2 [7]. In addition, the NC1 domain of the $\alpha 3(IV)$ chain has been identified as the autoantigen in Goodpasture's syndrome [26, 56]. Interestingly, human mRNAs for the $\alpha 3(IV)$ NC1 are present in three species, a full-length one and two truncated ones [25], although the possible relevance of this finding is unclear, as no major differences in the expression of the complete and truncated forms were found in kidneys of normal and Goodpasture-affected individuals [25]. Many mutations have been found in the gene encoding the $\alpha 5(IV)$ chain in the more common X-linked form of Alport's syndrome [46]. The disease is characterized by a progressive kidney condition and sensorineural deafness, demonstrating the essential role of the $\alpha 5(IV)$ chain of type IV collagen in the GBM and inner ear. In some families the Alport syndrome cosegregates with diffuse leiomyomatosis, a benign smooth muscle tumor diathesis. These patients have genomic deletions that disrupt the genes for the $\alpha 5(IV)$ and $\alpha 6(IV)$ chains, which are located in a head-to-head arrangement on Xq22 [24]. Thus these two type IV collagen chains may regulate smooth muscle differentiation and morphogenesis.

Laminin

Laminin is a major noncollagenous basement membrane-specific protein which is important as a structural component as well as for its role in many biological functions, including cell attachment, proliferation and differentiation [for reviews, see 1, 4, 8–11, 57]. Laminin is a heterotrimer, each molecule being composed of an A chain (300–400 kD), a B1 chain and a B2 chain (each 180–200 kD). The three chains assemble into a characteristic cross-shaped structure with three short arms and one long arm. Much of the biochemical and structural information is derived from laminin isolated from the basement membrane matrix-forming Engelbreth-Holm-Swarm (EHS) mouse tumor. Electron microscopy of EHS laminin reveals one of the short arms to be longer than the other two, about 48 nm vs. about 34 nm [58]. All three short arms have a terminal and a central globular domain separated by rod-like segments, and the 48-nm arm contains an additional globular structure near the concurrence of the three chains (fig. 4). The long arm of laminin is a flexible rod about 72 nm in length and 3 nm in width and contains a large terminal globule which is composed of five 4-nm subdomains [8, 57]. It has recently become evident that marked heterogeneity exists with respect to laminin structure, as nine distinct subunit chains and at least six laminin molecule isoforms have been identified, some of which deviate in shape from the classical laminin molecules. Thus some of the molecular heterogeneity and diverse functions of basement membranes are likely to be attributable to the various laminin isoforms.

Laminin chains can be divided into three groups in terms of their structural characteristics and occurrence in heterotrimers, A, B1 and B2 chains. The growing complexity of the laminin family has made it necessary to introduce a more comprehensive nomenclature for laminin chains and assembly forms [59]. According to this proposal, the nomenclature of A, B1 and B2 chain types should be replaced by α, β and γ chains with different isoforms of a given chain type identified by Arabic numbers. This new nomenclature would allow better for the designation of additional members of the laminin family. Furthermore, a simple nomenclature for naming the various laminin assembly forms, laminin 1, 2 etc. in chronological order of their discovery, has been proposed [59]. The structures and tissue distributions of the various laminin chains are summarized in table 1.

The EHS laminin chains have been designated as Ae, B1e and B2e (e for EHS) or α1, β1 and γ1, respectively. Complete primary structures for these three chains have been reported for man [60–62], the mouse [63–65] and *Drosophila* [66–69]. Two α-chain variants have been identified, Am (m for merosin) or α2 [70], which has been sequenced completely from man [71], and a 165-kD chain, termed At (t for truncated) or α3 [72] with either short or

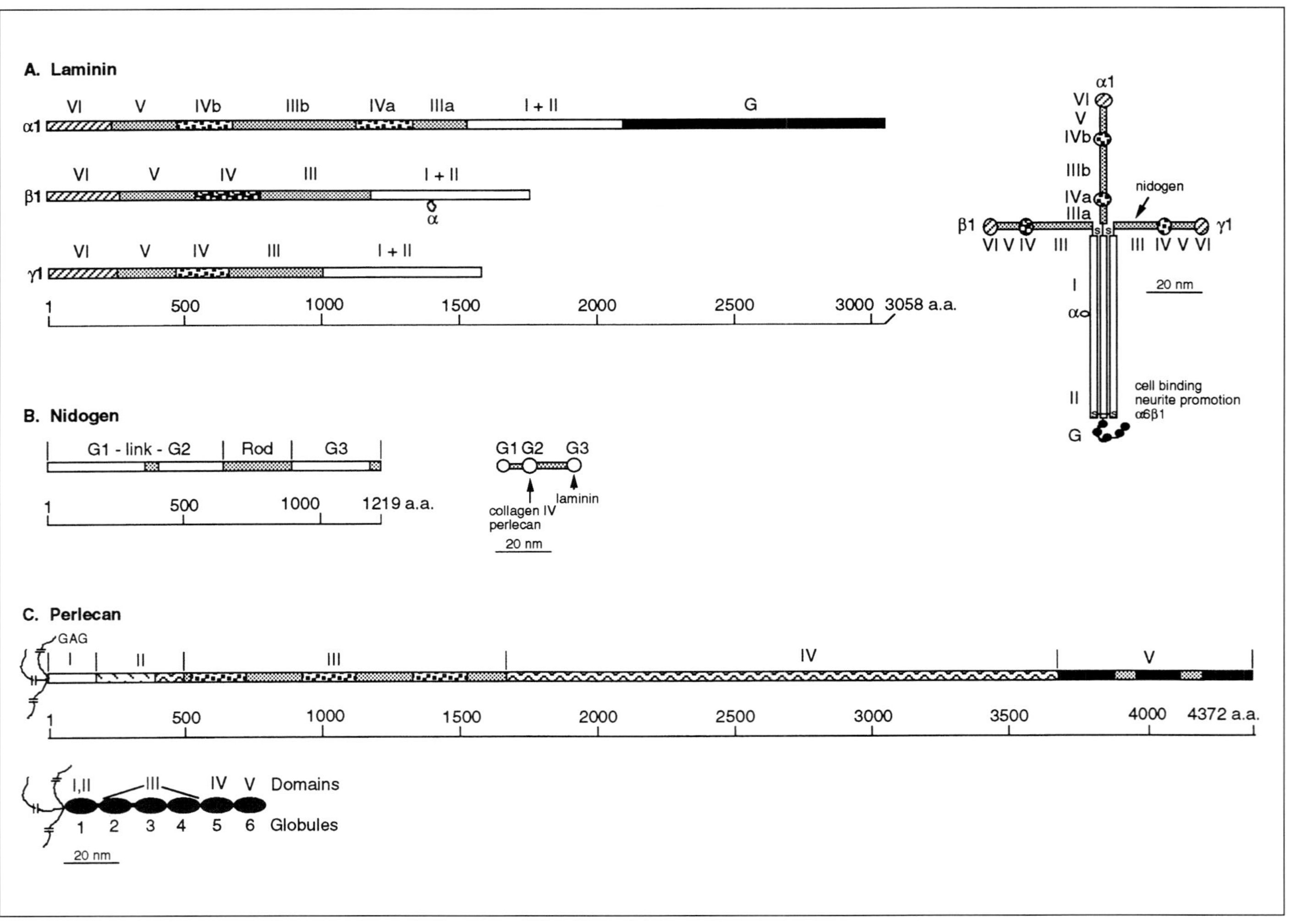

A. Laminin
α1
VI V IVb IIIb IVa IIIa I + II G
β1
VI V IV III I + II
α
γ1
VI V IV III I + II
1 500 1000 1500 2000 2500 3000 3058 a.a.
α1
VI V IVb IIIb IVa IIIa
nidogen
β1 III III γ1
VI V IV VI V IV
I
αo
II
cell binding
neurite promotion
α6β1
G
20 nm
B. Nidogen
G1 - link - G2 Rod G3
1 500 1000 1219 a.a.
G1 G2 G3
collagen IV laminin
perlecan
20 nm
C. Perlecan
GAG
I II III IV V
1 500 1000 1500 2000 2500 3000 3500 4000 4372 a.a.
I,II III IV V Domains
1 2 3 4 5 6 Globules
20 nm

long NH$_2$ termini [73]. The human α1 and α2 chains are 3058 and 3088 residues in length, respectively, while the shortest form of the α3 chain is only 1693 residues, lacking most of the short arm sequences. The first undisputable evidence of laminin heterogeneity came from the cloning of a variant rat B1 chain, termed B1s (s for synaptic) or β2 [74], subsequently also cloned from human [75]. A truncated form of the B1 chain, β3, also exists [72, 76]. A fourth B1 chain, β4, was detected by characterization of a short chicken eye cDNA clone encoding 198 amino acid residues with sequence homology with the β1 and β2 chains [77]. The complete primary structure of a truncated B2 chain homologue, B2t or γ2, has been reported for man [78] and the mouse [79], and found to exist possibly in two forms differing in the length of their COOH-terminal region. The human β1 and γ1 chains are 1765 and 1576

Fig. 4. Schematic illustrations of the multidomain structures of human laminin 1 (*A*), nidogen (*B*) and perlecan (*C*). Both individual polypeptide chains and electron micrography-derived schematic models of the isolated proteins are shown. *A* The three chains of laminin 1, α1, β1 and γ1, are shown with their distinct domains indicated by Roman numerals I–VI, α and G. The G domain contains five homologous repeats of about 200 residues, domains I and II contain repeated sequences that fold into α-helical conformation and form a coiled-coil structure of the three parallel chains, domain α represents an interruption in the above repeated sequence and domains III (a,b) and V are cysteine-rich repeats. Regions IV (a,b) and VI are globular domains and those showing homology are indicated with the same symbols. Sites involved in nidogen binding, cell binding, promotion of neurite outgrowth and binding of α6β1 integrin are indicated as well as a disulfide bond between the β1 and γ1 chains (S-S) and putative disulfide bonds (SS) occurring at the center of the cross-shaped laminin molecule. *B* Nidogen encompasses three globular domains G1–G3, a link region separating domains G1 and G2 and a rod-like domain separating G2 and G3. Cysteine-rich repeats are shown in gray while unique sequences are shown in white. Major binding regions for type IV collagen, perlecan and laminin are shown. *C* Amino acid sequences of perlecan can be defined into five domains indicated by Roman numerals I–V and they appear as 6 globules (1–6) in the visualized protein. Domain I is unique in sequence and contains attachment sites for glycosaminoglycan chains (GAG). Domain II contains two regions of homology to other proteins, namely the NH$_2$-terminal portion consists of four LDL receptor-like repeats and the COOH-terminal portion of one Ig-like repeat. Domain III contains four cysteine-rich repeats similar to those found in domains II and V of the laminin chains and three subdomains that are homologous with domains IVa and IVb of the laminin α chain. Domain IV contains 21 copies of Ig-like repeats while domain V contains three subdomains homologous to the repeats found in domain G of laminin and two regions of cysteine-rich repeats. The lengths of the different polypeptide chains are shown in terms of their amino acid residues (a.a.) and the schematic presentations of the molecules visualized by electron micrography are drawn approximately to scale. The polypeptide structures are based on references 63–65, 119 and 143, and other original data utilized in the figure are cited in the text and in reviews 9, 129 and 131.

residues in size, respectively. One or more of the truncated chains are thought to occur in the newly described laminin variants, kalinin [80], also known as epiligrin [81], and K-laminin [82], which is specific to epithelial cells. Kalinin has a Y-shaped appearance in electron microscopy with a long arm with a large globular domain at one end, and two short arms with just one globule on each of them. K-laminin is otherwise similar but its two short arms have doublet globular domains.

Full-length laminin chains have been found in four combinations in a number of tissues, namely $\alpha 1$-$\beta 1$-$\gamma 1$ (EHS-laminin or laminin 1), $\alpha 2$-$\beta 1$-$\gamma 1$ (merosin or laminin 2), $\alpha 1$-$\beta 2$-$\gamma 1$ (s-laminin or laminin 3), and $\alpha 2$-$\beta 2$-$\gamma 1$ (merosin/s-laminin or laminin 4) [83, 84]. The exact locations of the various laminin isoforms in tissues is not clear, but immunolocalization with chain-specific antibodies and localization of corresponding mRNAs by in situ hybridization and/or Northern blotting have been attempted. mRNAs for laminin chains of the EHS type have been found to be expressed in different proportions in mouse tissues, and the ratios of the different chains vary during development, possibly reflecting the existence of laminin isoforms [85–87]. Immunohistological findings indicate that the $\alpha 1$ and $\alpha 2$ chains and the $\beta 1$ and $\beta 2$ chains are often mutually exclusive in their tissue distribution [83, 88, 89]. The $\beta 1$ and $\gamma 1$ chains have a wide tissue distribution, the latter possibly being the most abundant laminin chain [78, 83, 88]. Northern and in situ hybridization studies comparing the expression of the $\beta 1$ and $\beta 2$ chain mRNAs in fetal tissues indicate that many cell types expressing the $\beta 1$ chain also expressed the $\beta 2$ chain, but in certain locations, such as kidney glomeruli, hepatocytes, the epidermis and the periventricular mantle zone in the brain, the $\beta 2$ chain is strongly expressed, while the $\beta 1$ chain is not [75]. The $\beta 2$ is also concentrated in the synapses of motor neurons [74]. The $\alpha 1$ chain has the most restricted tissue distribution, being detected mainly in epithelial basement membranes, while endothelial, mesenchymal and smooth muscle cells express at most only low levels of it [90–92]. $\alpha 1$ mRNA expression is detected only in the kidney and brain tissues of newborn human infants [62]. The epithelial cells appear to be the main source of the $\alpha 1$ chain in the developing mouse kidney and immunostaining of adult mouse kidney detects the $\alpha 1$ chain in the GBM and tubular basement membrane [88, 90]. The $\alpha 2$ chain mRNA has a wide distribution and is primarily expressed by cells of mesenchymal origin, while epithelial and endothelial cells appear to be negative for this chain [71]. Immunostaining of the human kidney shows the glomerular and tubular basement membranes to be negative for the $\alpha 2$ chain [88]. Regarding the occurrence of the $\beta 1$ and $\beta 2$ chains in the kidney, the former is found in the tubular basement membrane and the mesangium and the latter only in the GBM , while the $\gamma 1$ chain is present in both locations [88].

The recently described laminin isoforms kalinin and K-laminin are thought to have chain compositions of $\alpha3$-$\beta3$-$\gamma2$ and $\alpha3$-$\beta1$-$\gamma2$, respectively, and have been located immunologically in the basement membranes of the epithelial-mesenchymal junctions in the skin, lung, gut and amnion [80]. In situ hybridization experiments with the $\gamma2$ chain of kalinin have demonstrated that expression is confined to the epithelial cells [78]. No expression of these laminin isoforms has been detected in the GBM.

Laminin is a mosaic protein containing several modular units also found in some other extracellular matrix proteins [8, 57]. Figure 4 outlines the general structure and various domains of laminin as obtained from sequence information and structural and functional analyses. The sequence information indicates that the large globular domain (G) on the long arm of laminin is formed entirely of the COOH-terminal end of the α chain. Here the α chain contains five internally homologous repeats of about 200 residues. The rod-like region of the long arm of laminin is jointly formed by all three chains from their domains I and II, which encompass about 600 residues. These domains have little sequence homology between the three chains, but they all contain heptad repeats that fold into α-helical conformation [93, 94]. The three chains align in parallel to each other and form a coiled-coil structure, the COOH-terminal ends of the β and γ chains being joined by a disulfide bridge and at least three additional disulfide bridges existing between the various chains near the center of the cross. The coiled-coil region contains several irregularities, the largest being a cysteine-rich domain, α, of about 30 residues in the β chain. Experimental data and calculations of ionic interaction scores between heptad repeat regions in the various laminin chains suggest that the β and γ chains are required for the coiled-coil structures, homodimers of β and γ chains probably being unstable. The NH_2-terminal 1100–1300 residues of the chains of the three classes (less in the truncated chains) show high structural similarity, including distinct globular regions and rod-like cysteine-rich sequences. Each short arm is formed by the NH_2-terminal portions of one of the three chains and consists of alternating cysteine-rich repeats (domains III and V) and globular domains (IV and VI). The cysteine-rich sequences are homologous to the epidermal growth factor motif. The NH_2-terminal globular domains of all three chains are homologous and the inner globular domains of the α and γ chains share sequence homology. Domains III and IV occur twice in the α chain, domains IVa and IVb corresponding to the two inner globules seen in the α chain-derived short arm.

The laminin chains contain many potential sites for asparagine-linked glycosylation, most of which occur within the long arm sequences, and they are about twice as frequent on the α chain as on the β and γ chains. It has been estimated that about 40 of the putative sites are glycosylated and processed in the Golgi apparatus, the degree of glycosylation being 12–15% (w/w) in earlier reports and 25–30% (w/w) in more recent ones [95–98]. N-glycosylation does not seem to affect the biosynthetic assembly of laminin chains into trimeric molecules [99], but it may affect the interaction of laminin with its receptors, as has been shown for the $\alpha6/\beta1$ integrin [100]. As with collagens, it is evident that various laminin chains are synthesized in the same tissue and even the same cell. Thus the assembly of laminin molecules from the proper chain combinations must be dictated by the association of the correct set of three chains. Biosynthetic and in vitro folding experiments suggest that disulfide-linked heterodimers of β and γ chains occur as intermediates during folding, and that the α chain may be added at a later stage of folding [101, 102]. Initial random association of the individual laminin chains has also been reported, however [99]. Following secretion, laminin molecules are thought to self-assemble into complex polymers. In vitro experiments indicate that laminin molecules can self-assemble into a network structure in the presence of Ca^{2+} in a concentration and temperature-dependent manner [51]. Laminin molecules interact with each other via their terminal globular domains, and electron microscopy has shown that dimers, trimers and oligomers form through these interactions [51].

Examination of in vitro-generated laminin aggregates and basement membranes from which type IV collagen has been removed by digestion with bacterial collagenase indicate that laminin forms an independent network in basement membranes [51]. Expression of laminin $\beta1$ and $\gamma1$ chains can already be detected at the two-cell stage of mouse development, clearly preceding the expression of type IV collagen and nidogen [103]. These results suggest that laminin may make up the basement membrane scaffold in newly formed basement membranes during their development, for example, and type IV collagen is added only at later stages. Furthermore, laminin interacts with other basement membrane components to generate the correct basement membrane architecture, and the formation of stable equimolar complexes is particularly well established with nidogen, which appears to be critical for joining the two types of network (see below).

Besides its structural role, laminin is implicated in many cellular processes. A voluminous literature describes the various biological activities of laminin, including cell adhesion and migration, growth and differentiation, promotion of neurite outgrowth and receptor binding [1, 4, 8–11, 57, 104]. Functional domains have been mapped by assessing the effects of proteolytic fragments of laminin, usually of the EHS type, and synthetic peptides in in vitro cell or tissue culture systems. Some of these results may lack any biological significance, however, as the activities of native laminin are sometimes different from those of idolated subunits. Laminin is a major cell binding component in basement membranes, and the interactions of laminin with cells have been found to be mediated by several cellular receptors. At least eight integrins have been reported to bind EHS laminin: $\alpha1\beta1$, $\alpha2\beta1$, $\alpha3\beta1$, $\alpha6\beta1$, $\alpha7\beta1$, $\alpha v\beta3$, $\alpha IIb\beta3$ and possibly $\alpha6\beta4$ [8–11, 105]. Laminin appears to be the only known ligand for $\alpha6\beta1$ and $\alpha7\beta1$ integrins in several cell types, while most of the others bind several ligands. The $\alpha6\beta1$ integrin interacts with kalinin and merosin in addition to EHS-laminin and the $\alpha3\beta1$ integrin with kalinin [81, 106]. The major cell binding site and neurite-promoting site have been located at the end of the long arm of laminin. An Arg-Gly-Asp-independent recognition site for $\alpha3\beta1$ and $\alpha6\beta1$ has been located at the tip of the long arm [107, 108]. Additional recognition sites exist in the short arms. Several nonintegrin receptors of varying size have also been found to bind laminin, including several that bind carbohydrates. A 67-kD high-affinity laminin binding protein has been found in many cell types [105] Expression of the mRNA for this receptor has been found to precede laminin and type IV collagen expression in kidney development, which may indicate that it plays a role in the process [91]. Characterization of this receptor has produced controversial results, however, as cDNA clones for it encode a cytoplasmic component of the translation machinery [109].

Furthermore, direct evidence for the essential role of kalinin in epithelial attachment has been obtained by studies on junctional epidermolysis bullosa. This autosomal recessive disorder is characterized by blister formation within the dermal-epidermal basement membrane, and recently mutations in the $\gamma2$ chain of kalinin were identified in 2 patients [110].

Nidogen/Entactin

Nidogen [111], also known as entactin [112], is an abundant glycoprotein found in all basement membranes. Its expression during embryogenesis appears at the 8- to 16-cell morula stages, thus closely following the appearance of laminin [103]. Immunoelectron microscopy demonstrates nidogen throughout the entire thickness of the GBM at both developing and mature stages, while labeling in Bowman's capsule decreases during maturation [113].

Nidogen consists of a single 148-kD polypeptide containing 5% carbohydrate moietes, including both N- and O-linked oligosaccharides, and tyrosine-O-sulfate [114, 115]. Complete cDNA-derived sequences have been established for the human [116] and mouse [117, 118], indicating that the mature polypeptide consists of 1219 and 1217 residues, respectively, with 85% sequence identity between the two species (table 1). The human polypeptide consists of a 28-residue signal peptide, an 643-residue NH_2-terminal sequence with one epidermal growth factor (EGF)-type cysteine-rich repeat near the middle, a 248-residue sequence consisting of five EGF repeats, and a 328-residue COOH-terminal sequence that terminates in an EGF repeat and contains repeating motifs showing homology to the low-density lipoprotein receptor and the EGF precursor protein (fig. 4). Furthermore, deduced amino acid sequences reveal two putative N-linked carbohydrate acceptor sites and two consensus sequences for tyrosine sulfatation.

Electron microscopy of nidogen has shown it to be a dumb-bell-shaped protein with a large and a small globular domain connected by a 17-nm rod-like segment [1, 4, 12]. This finding represents authentic nidogen extracted from the EHS tumor under denaturing isolation conditions. Recombinant mouse nidogen has recently become available from stably transfected mammalian cells [119], and since it can be purified without exposure to denaturing agents it represents a more physiological material for use in structural and functional studies. Characterization of this recombinant nidogen resulted in a modified domain model consisting of three globular domains, G1–3, domains G1 and G2 being connected by a thin segment and G2 and G3 by a rod [119]. The correlation between the morphological structure and amino acid sequence data is shown in figure 4. Of the major basement membrane proteins, nidogen is the most protease-sensitive, by virtue of which it is thought to play a key role in the remodeling of basement membranes [1, 4, 12]. Several protease-sensitive sites occur along the molecule, the link between the two NH_2-terminal globular domains G1 and G2 being particularly easily subject to degradation.

Nidogen is found in tissues in close association with laminin, the two being extractable in a 1:1 molar ratio from several basement membrane sources. EHS tumor nidogen binds to one of the short arms of laminin in a noncovalent fashion to form a highly stable complex which requires guanidine denaturation for dissociation. The binding has been mapped to occur between the COOH-terminal globular domain, G3, of both authentic and recombinant nidogen and a sequence encompassing one EGF-like motif, repeat No. 4, in the domain III of the laminin γ1 chain [120–122]. Nidogen also binds to type IV collagen, but in a more complex manner, some ambiguity existing in the locations of the binding sites in the two molecules. The collagen binding site in nondenatured recombinant nidogen has been mapped to within or close to its G2 domain [119], while several putative binding regions have been located in the triple helical domain of the collagen [123]. Thus the laminin and type IV collagen binding sites in nidogen appear to be distinct and are separated by the rod-like domain. Kd values of <1 nM were observed for the interactions of recombinant nidogen with both laminin and type IV collagen, the binding activity thus being about 10-fold higher than that observed with guanidine-HCl-isolated nidogen [119]. Furthermore, nidogen binds to heparan sulfate proteoglycan, the binding occurring to the core protein and not to the heparan sulfate chains [124]. Thus the multiple binding activities of nidogen indicate that it probably serves as an important link in the structural organization of the various components of basement membranes (see below). Nidogen has also been found to bind to fibronectin and fibrinogen, proteins not specific to basement membranes but sometimes found close to these structures [125].

Nidogen contains two EGF-like repeats in the rod-like domain that bear homology to various calcium binding proteins and two additional putative binding sites of the EF motif in or near the G1 domain. Experiments utilizing full-length recombinant nidogen and recombinant fragments detected Ca^{2+} binding in both the rod-like domain and G1, but this is not required for the interactions of nidogen with laminin, type IV collagen and proteoglycan [126]. The physiological significance of the Ca^{2+} binding property of nidogen is therefore not clear at present.

In addition to its role in extracellular matrix assembly, several other less well established activities of nidogen have been described, including cell adhesion. Nidogen contains one Arg-Gly-Asp sequence and several variant sequences which may be involved in cell binding, although results regarding this aspect remain unclear [126]. Nidogen may also play a role in hemostasis and wound healing, and enhance chemotaxis and phagocytosis [125, 126].

Proteoglycans

Proteoglycans, which are present in all extracellular matrices [127, 128], consist of several types of core protein with covalently attached glycosamino-glycan side chains, usually on serine residues, that show some diversity in size and chemical composition. The proteoglycans found in basement membranes are unrelated to the many others found elsewhere in the extracellular matrix and on cell surfaces. The first demonstrations of proteoglycans in basement membranes came from histological studies suggesting the presence of anionic sites in the GBM, later enzymatically identified as heparan sulfate [129, 130]. Subsequently the essential role of heparan sulfate proteoglycan (HSPG) in glomerular filtration was demonstrated, as its removal by heparitinase increased the permeability of the GBM [131]. Heparan sulfate is the most abundant glycosaminoglycan in human, bovine and rat GBM, but it also contains small amounts of chondroitin sulfate and hyaluronic acid [127, 128]. Comparison of the glycosaminoglycan compositions of the various basement membranes has indicated clear differences in the ratio of heparan sulfate to chondroitin sulfate, and considerable heterogeneity also exists regarding the core proteins.

The best studied and most abundant basement membrane proteoglycan is perlecan [128], a very large, low-density HSPG present in all basement membranes. Perlecan isolated from the EHS tumor is over 600 kD in size and contains a single 400–450 kD polypeptide with two or three 65-kD glycos-aminoglycan chains [132]. Immunologically similar molecules have also been found in other basement membranes, such as the GBM, and in other species, the size varying in the range 400–520 kD [132–136]. The major glomerular HSGP was initially characterized as being 130 kD in size with an 18-kD core protein, but several larger glomerular core protein sizes ranging up to 250 kD were subsequently characterized [128]. Immunoprecipitation of rat glomerular material using antibodies directed against the large EHS HSPG detected core proteins varying in size between 150 and 400 kD, the predominant species being 250 kD [137]. It was thus suggested that the 400 kD core protein in glomeruli represents a precursor form that undergoes proteolytic processing which produces smaller molecular weight HSPGs than are usually found in the EHS tumor. Furthermore, perlecan has been found to be sensitive to proteolysis, which may in part explain some of the heterogeneity found. All in all, the results regarding the many smaller core proteins found in GBM and other basement membranes are complex and the basis for the structural heterogeneity has not been fully resolved. Some of the small core proteins share antigenic determinants with perlecan, thus possibly representing precursor-product relationships, while others are immunologically distinct from it, thus representing separate gene products or proteolytic fragments of perlecan lacking the antigenic epitopes [128].

Perlecan is synthesized by a wide variety of cultured cells, including fibroblasts, which do not produce basement membranes [128]. Interestingly, dermal fibroblasts also produce perlecan in vivo and deposit it in the surrounding matrix, indicating that fibroblast expression is not merely an artifact of tissue culture [135]. Nevertheless, the major sites of perlecan accumulation in vivo are basement membranes. Electron microscopy indicates that perlecan is folded into an elongated structure about 80 nm long with a tandem array usually of six globular domains of highly variable diameter, sometimes seen to be separated by short rod-like connections [138]. Hence the name perlecan. Furthermore, the molecule generally contains three 100–170 nm heparan sulfate side chains asymmetrically attached to one end (fig. 4). Immunostaining has enabled perlecan to be located in basement membranes throughout their thickness [128], and immunogold microscopy using monoclonal antibodies specific to distinct epitopes of the mouse perlecan core protein suggest that the molecule may adopt a specific orientation in basement membranes [135]. The molecules appear to span the entire membrane, with the glycosaminogly-can end facing the cellular side and the COOH-end facing the underlying matrix [135]. The estimated length of perlecan, 80 nm, is compatible with such a proposal in the case of most basement membranes.

cDNA clones for mouse perlecan predict a 3707-amino acid residue polypeptide with a calculated molecular weight of 396 kD [138, 139]. The human polypeptide is larger, 4372 residues, with a molecular weight of 467 kD, due to a substantially longer repeat sequence [140] (table 1). In both species the polypeptide contains an additional 21-residue signal peptide. The sequences indicate that perlecan is a mosaic protein consisting of five domains, four of which represent sequence motifs found in other proteins. The following multidomain structure was found in the human protein (fig. 4): domain I, a unique NH$_2$-terminal 172-residue sequence with three putative Ser-Gly-Asp glycosaminoglycan attachment sites; domain II, a 210-residue sequence with four LDL receptor-like repeats containing six conserved cysteines in each and one Ig-like repeat; domain III, a 1172-residue sequence resembling the short arms of laminin; domain IV, a 2010-residue sequence containing 21 Ig-like repeats, and domain V, a 705-residue sequence resembling the COOH-terminal globular domain of the laminin α chain. The human and mouse sequences are highly conserved with the exception of domain IV, which contains only 14 Ig-like repeats in the mouse [138–141]. The difference may be due to alternative splicing of primary transcripts or actual size differences in the proteins. Domain III contains four cysteine-rich repeats similar to those found in domains II and V of the short arms of the laminin chains, and these can be predicted to fold into rod-like structures. These subdomains separate three more subdomains that are highly similar to the globular domains IVa and IVb of the laminin α chain. The highly repetitive domain IV of perlecan is most homologous with the Ig repeats in the neural adhesion molecule, the sequence conservation including two conserved cysteines in each repeat. The COOH-terminal domain V contains three subdomains homologous to the five COOH-terminal repeats in domain G of the laminin α chain. The three subdomains are separated by two sequences, each of which consists of two copies of EGF-like repeats. The human chain contains 10 and the mouse 12 putative N-linked glycosylation sites scattered among all the domains, and several of them are conserved between the two species.

Figure 4 shows the correlation between the primary structure of perlecan and the multiglobular molecule when visualized by electron microscopy. Domains I and II jointly are likely to form the globule containing the glycosaminoglycan chains. Domain III is predited to fold in the same way as the homologous short arm region of the laminin α chain, namely into three globules separated by rods. The fifth globule is predicted to be formed by domain IV and the sixth by domain V, by an analogy to the COOH-terminal globule of the laminin α chain. The structures of the various globules are likely to be stabilized by disulfide bonds [128].

The most notable feature of the biosynthesis of perlecan is the addition of the heparan sulfate side chains, but chondroitin sulfate side chains have also been observed [128]. The molecule is also subject to additions of both N- and O-linked oligosaccharides. The nature and extent of the proteolytic processing thought to be included in the posttranslational processing of perlecan has not yet been fully characterized, and perlecan may also be subject to fatty acylation [128].

The function of the anionically charged heparan sulfate side chains of perlecan in the charge selectivity of GBM is well established [128, 131]. The loss of proteoglycans from the GBM observed in a number of diseases is associated with proteinuria. The glomerular endothelial and epithelial cells contain a thick anionic cell surface coat of sialoproteins, which probably contribute to the charge selectivity [14]. Proteoglycans of various tissue locations, including basement membranes, have also been found to serve as major binders of growth factors and cytokines [142]. The heparan sulfate side chains in particular have been found to interact with several growth factors, and at least one growth factor, transforming growth factor β, binds to the core protein. Other functions are likely to be attributed to the perlecan core protein, which is highly versatile in sequence terms [128]. Perlecan may be involved in basement membrane assembly, as it enhances the in vitro formation of basement membranes [143], and the two regions of perlecan that are homologous with laminin are thought to be involved in the putative structural function of perlecan in basement membranes. Purified EHS proteoglycan can self-assemble into dimers, and to a lesser extent into oligomers, via interactions between the COOH-terminal ends of the molecules [51]. Other properties include cell binding, calcium binding, regulation of serine protease activity, and growth- and neurite-promoting activities.

Small high-density heparan sulfate proteoglycans have also been identified in basement membranes, some of which are likely to represent degradation products of perlecan or cell surface proteoglycans released into the matrix [128]. A 130-kD proteoglycan with four heparan sulfate chains was isolated from the EHS tumor [144], and a similar molecule has also been found in the GBM [145]. These molecules share immunological determinants with perlecan and it has been suggested that they may represent its degradation products [136, 146]. Thus sequence information is needed for conclusive results on the identity of the various heparan sulfate proteoglycans in basement membranes. It is likely that both genetic differences and differences in posttranslational modifications contribute to the observed heterogeneity in basement membrane HSPGs.

Other proteoglycans identified in basement membranes include a high-density proteoglycan with a core size varying between 21 and 34 kD having both heparan sulfate and chondroitin sulfate side chains [147]. The peptide map of the core protein was different from perlecan suggesting a separate gene product. A proteoglycan with a 160-kD core protein and 13-22 chondroitin sulfate side chains has been found in most basement membranes [148]. Immunostaining for this component was absent from the GBM of the adult rat kidney, however, although it was present in the mesangia, Bowman's capsule and tubular basement membranes. Furthermore, a large dermatan sulfate proteoglycan with complement inhibitory activity has been found to be synthesized by cultured rat glomerular epithelial cells [149].

Biosynthetic experiments using isolated glomeruli maintained in tissue culture or cultured glomerular epithelial cells showed synthesis of both heparan sulfate and chondroitin sulfate proteoglycans [128], although the majority of the proteoglycans synthesized are of the chondroitin sulfate type under in vitro culturing conditions [128]. The opposite results are obtained when synthesis in vivo is investigated. In vivo SO_4 labelling and in situ fixation of the newly synthesized proteins indicates that 85% of the proteoglycans synthesized by rat glomeruli are of the heparan sulfate type [150].

Other Components
A number of other proteins have been identified in basement membranes, but in most cases little information exists on their structure and/or function. Fibronectin is a well-characterized protein that exists both as a widely distributed extracellular matrix component and in a circulating form [126]. The presence of fibronectin in the GBM has been shown immunohistochemically [14]. The possible function of fibronectin in basement membranes is not known, however. It is not a typical basement membrane product, as it may represent circulating molecules trapped in the membrane. BM-40 is a protein found in the EHS tumor basement membrane [151], but it is not specific to basement membranes, as the same protein is also called osteonectin due to its presence in bone matrix and its function of linking mineral to collagen [152] and SPARC (soluble protein that is acidic and rich in cysteine), identified as a major endothelial and endodermal cell product [153]. All in all, it is expressed in a large number of extracellular matrices, including several basement membranes [126]. The EHS tumor and Reichert's basement membranes are rich in BM-40, while its presence in basement membranes such as the GBM has been more difficult to visualize, possibly due to masking of the epitope [126]. BM-40 has been cloned from several species and found to consist of 285–287 residues and have high-affinity calcium binding properties. Its role in basement membranes is not clear, however. A 72–80 kD glycoprotein called bamin has recently been isolated from the EHS tumor and detected immunohistologically in the rat GBM, although not in other renal basement membranes [154].

Synthesis of GBM Components during Development

Many detailed and sophisticated studies have been performed on kidney development. Renal glomerular epithelial cells, Bowman's capsule and the tubules originate from mesenchymal cells induced to undergo epithelial differentiation by a branch of the uretic bud. The differentiation includes comma-shaped, S-shaped body and capillary loop stages. The ensuing primitive nephron is invaginated by mesenchymal cells that convert into the glomerular endothelium [103]. Electron microscopic, biosynthetic and immunohisto-chemical studies suggest that at the early stages of development both the endothelial and epithelial cells synthesize a basement membrane that fuses at later stages to give rise to the GBM [3, 155]. Molecular changes in the composition of the GBM occur during the maturation process and it becomes progressively less permeable to macromolecules [155]. Studies on the ontogenesis of rat GBM using immunoelectron microscopy indicate labeling of both endothelial and epithelial basement membranes for the $\alpha1(IV)$ and $\alpha2(IV)$ chains of type IV collagen and nidogen [113]. The three polypeptides are present throughout the GBM after fusion of the two basement membranes, but the labeling for the $\alpha1(IV)$ and $\alpha2(IV)$ collagen chains decreases during maturation and shifts towards the endothelial side, while nidogen continues to be located centrally. The $\alpha3(IV)$ labeling was found to be located centrally in the GBM and clearly increased during maturation. Different patterns of maturation were observed between the GBM and Bowman's capsule, as labeling of the latter for the $\alpha1(IV)$ and $\alpha2(IV)$ chains remained high throughout maturation. In vivo labeling of the GBM using anti-EHS laminin IgG suggests that after fusion of the basement membranes new material is added predominantly by the epithelium [155]. Laminin $\alpha1$, $\beta1$ and $\gamma1$ chain expression during kidney development has been studied in detail [103]. At the initial stages the kidney mesenchymal cells express predominantly the $\beta1$ and $\gamma1$ chains, and expression of the $\alpha1$ chain is very low. Subsequently, $\alpha1$ chain expression increases markedly in company with epithelial cell development, and $\beta1$ and $\gamma1$ chain expression continues at significant levels. In situ hybridizations indicate that laminin $\alpha1$ mRNA is restricted to the epithelial cells in the developing tubules, while $\beta1$ and $\gamma2$ chain mRNAs are produced by both epithelial and mesenchymal cells [103, 156]. As $\alpha1$ chain expression accompanies the formation of a polarized epithelium, it has been postulated that this chain is important for cell polarization [103]. Type IV collagen $\alpha1(IV)$ chain mRNA has also been found in both epithelial and mesenchymal cells in kidney, but nidogen mRNA only in mesenchymal cells, all other embryonic basement membranes were also such that nidogen was produced by mesenchymal and endothelial cells but not by the epithelium [156]. Collectively, the results indicate that basement membranes are assembled from components produced by more than one cell type. Differential expression of a small HSPG and a chondroitin sulfate proteoglycan [148] has also been detected during organogenesis, the former being identified in all kidney basement membranes throughout development and the latter up to the formation of the glomeruli, after which its expression decreased in the capillary basement membrane but continued in the mesangium and Bowman's capsule [157]. The mature GBM is unique in lacking the chondroitin sulfate proteoglycan, but this appears to be needed during kidney development. Furthermore, many results indicate that the GBM and the mesangial matrix are similar in composition, although not identical. Evaluations of collagen synthesis in cultured glomerular epithelial cells and mesangial cells have shown that type IV is the major collagen synthesized by the former cells while the latter synthesize almost equal amounts of types I and IV and small amounts of type III [158]. Furthermore, the mesangium contains type VI collagen.

Supramolecular Organization of GBM Components

The interactions between the major basement membrane components have been characterized on many occasions, and specific regions with high binding activities between the various components have been identified. The basic structure of basement membranes is generated by self-assembly of the type IV collagen and laminin molecules into separate networks [51]. The type IV collagen molecules form complex associations stabilized by extensive cross-linking, resulting in an irreversible network. Thus the collagen molecules are virtually unextractable. The basement membranes contain more than one assembly form of type IV collagen molecules, which dimerize through their COOH-terminal globular domains, mainly to form dimers with identical isoforms (homodimers), although heterodimerization apparently also occurs [7, 51]. Laminin polymerization is a reversible process, and stable laminin-nidogen complexes can be extracted from basement membranes using chelating agents. Results obtained with recombinant nidogen support a crucial structural role for nidogen in basement membrane organization and stabilization [159]. The significance of the type IV collagen and laminin isoforms for the structural integrity and other activities of basement membranes is not yet understood. Nidogen is capable of binding to type IV collagen, laminin and perlecan, and it also effectively mediates ternary complex formation between type IV collagen and laminin and between laminin and perlecan [159]. The type IV collagen and laminin networks are thought to form an interwoven mesh in which nidogen molecules may bridge the two types of polymer. The molecular sieving property of the collagen and laminin network structures is probably not sufficient, and anchorage of heparan sulfate molecules into the basic networks may be needed to generate the required filtration barrier [51, 160]. It is obvious that not all the components of basement membranes have yet been identified, and the precise supramolecular organization of the known components is not clear. Further work is therefore needed to understand the in vivo organization of basement membranes.

Acknowledgements

The author gratefully thanks Karl Tryggvason (University of Oulu, Finland) for critical reading of the manuscript and Auli Kinnunen for expert secretarial assistance. Supported by grants from the Research Council for Medicine within the Academy of Finland and the Sigrid Juselius Foundation.

References

1   Timpl R, Dziadek M: Structure, development, and molecular pathology of basement membranes. Int Rev Exp Pathol 1986;29:1–112.
2   Rohrbach DH, Timpl R: Molecular and cellular aspects of basement membranes; in Rohrbach DH, Timpl R (eds): Molecular and Cellular Aspects of Basement Membranes. San Diego, Academic Press, 1993, pp 1–448.
3   Martinez-Hernandez A, Amenta PS: The basement membrane in pathology. Lab Invest 1983;48: 656–677.
4   Timpl R: Structure and biological activity of basement membrane proteins. Eur J Biochem 1989; 180:487–502.
5   Tryggvason K: Biochemistry and genetic diseases of glomerular basement membrane. Semin Nephrol 1993;13:447–456.
6   Hudson BG, Wieslander J, Wisdom BJ, Noelken ME: Biology of disease. Goodpasture syndrome: Molecular architecture and function of basement membrane antigen. Lab Invest 1989;61:256–269.
7   Hudson GB, Reeders ST, Tryggvason K: Type IV collagen: Structure, gene organization, and role in human diseases. Molecular basis of Goodpasture and Alport syndromes and diffuse leiomyomatosis. J Biol Chem 1933;268:26033–26036.
8   Beck K, Hunter I, Engel J: Structure and function of laminin: Anatomy of a multidomain glycoprotein. FASEB J 1990;4:148–160.
9   Engel J: Laminins and other strange proteins. Biochemistry 1992;31:10643–10651.
10 Engvall E: Laminin variants: Why, where and when? Kidney Int 1993;43:2–6.
11 Tryggvason K: The laminin family. Curr Opin Cell Biol 1993;5:877–882.
12 Chung AE, Durkin ME: Entactin: Structure and function. Am J Res Cell Mol Biol 1990;3:275–282.
13 Dunnill M, Halley W: Some observations on the quantitative anatomy of the kidney. J Pathol 1973; 110:113–121.
14 Kasinath BS, Kanwar YS: Glomerular basement membrane: Biology and physiology; in Rohrbach DH, Timpl R (eds): Molecular and Cellular Aspects of Basement Membranes. San Diego, Academic Press, 1993, pp 89–106.
15 Takanami H, Naramoto A, Shigematsu H, Ohno S: Ultrastructure of glomerular basement membrane by quick-freeze and deep-etch methods. Kidney Int 1991;39:659–664.
16 Rodewald R, Karnovsky MJ: Porous substructure of the glomerular slit diaphragm in the rat and mouse. J Cell Biol 1974;60:423–433.

17   Goldberg M, Escaig-Haye F: Is the lamina lucida of the basement membrane a fixation artifact? Eur J Cell Biol 1986;42:365–368.

18   Caulfield JP, Farquhar MG: Loss of anionic sites from the glomerular basement membrane in aminonucleoside nephrosis. Lab Invest 1978;39:505–512.

19   Brenner BM, Hostetter TH, Humes HD: Molecular basis of proteinuria of glomerular origin. N Engl J Med 1978;298:826–833.

20   Tryggvason K, Zhou J, Hostikka SL: Alport syndrome and other inherited basement membrane disorders; in Rohrbach DH, Timpl R (eds): Molecular and Cellular Aspects of Basement Membranes. San Diego, Academic Press, 1993, pp 421–437.

21   Rohrbach DH, Murrah VA: Molecular aspects of basement membrane pathology; in Rohrbach DH, Timpl R (eds): Molecular and Cellular Aspects of Basement Membranes. San Diego, Academic Press, 1993, pp 385–419.

22   Mayne R, Brewton RG: New members of the collagen superfamily. Curr Opin Cell Biol 1993;5: 883–890.

23   Kivirikko KI: Collagens and their abnormalities in a wide spectrum of diseases. Ann Med 1993;25: 113–126.

24   Zhou J, Mochizuki T, Smeets H, Antignac C, Laurila P, de Paepe A, Tryggvason K, Reeders ST: Deletion of the paired α5(IV) and α6(IV) collagen genes in inherited smooth muscle tumors. Science 1993;261:1167–1169.

25   Bernal D, Quinones S, Saus J: The human mRNA encoding the Goodpasture antigen is alternatively spliced. J Biol Chem 1993;268:12090–12094.

26   Saus J, Wieslander J, Langeweld JPM, Quinones S, Hudson PG: Identification of the Goodpasture antigens as the α3(IV) chain of collagen IV. J Biol Chem 1988;263:13374–13380.

27   Johansson C, Butkowski R, Wieslander J: The structural organization of type IV collagen. Identification of three NC1 populations in the glomerular basement membrane. J Biol Chem 1992;267: 24533–24537.

28   Kleppel MM, Fan WW, Cheong HI, Michael AF: Evidence for separate networks of classical and novel basement membrane collagen. Characterization of α3(IV)-Alport antigen heterodimer. J Biol Chem 1992;267:4137–4142.

29   Sariola H, Hostikka SL, Lukkarila S, Tryggvason K: Distribution of Type IV collagen α1, α2 and α5 chains in human tissues; in Tryggvason K (ed): Molecular pathology and genetics of Alport syndrome. Contrib Nephrol. Basel, Karger, 1995, vol 117, pp 130–141.

30   Brazel D, Oberbäumer I, Dieringer H, Babel W, Glanville RW, Deutzmann R, Kühn K: Completion of the amino acid sequence of the α1 chain of human basement membrane collagen (type IV) revealed 21 non-triplet interruptions located within the collagenous domain. Eur J Biochem 1987; 168:529–536.

31   Soininen R, Haka-Risku T, Prockop DJ, Tryggvason K: Complete primary structure of the α1-chain of human basement membrane (type IV) collagen. FEBS Lett 1987;225:188–194.

32   Muthukumaran G, Blumberg B, Kurkinen M: The complete primary structure for the α1-chain of mouse collagen IV. Differential evolution of collagen IV domains. J Biol Chem 1989;264:6310–6317.

33   Blumberg B, MacKrell AJ, Fessler JH: *Drosophila* basement membrane procollagen α1(IV). II. Complete cDNA sequence, genomic structure, and general implications for supramolecular assemblies. J Biol Chem 1988;263:18328–18337.

34   Exposito J-Y, D'Alessio M, DiLiberto M, Ramirez F: Complete primary structure of a sea urchin type IV collagen α chain and analyses of the 5' end of its gene. J Biol Chem 1993;268:5249–5254.

35   Guo X, Johnsson JJ, Kramer JM: Embryonic lethality caused by mutations in basement membrane collagen of *C. elegans*. Nature 1991;349:707–709.

36   Hostikka SL, Tryggvason K: The complete primary structure of the α2 chain of human type IV collagen and comparison with the α1(IV) chain. J Biol Chem 1988;263:19488–19493.

37   Saus J, Quinones S, MacKrell A, Blumberg B, Muthukumaran G, Pihlajaniemi T, Kurkinen M: The complete primary structure of mouse α2(IV) collagen. Alignment with mouse α1(IV) collagen. J Biol Chem 1989;264:6318–6324.

38   Pettitt J, Kingston IP: The complete primary structure of a nematode α2(IV) collagen and the partial structural organization of its gene. J Biol Chem 1991;266:16149–16156.

39 Sibley MH, Johnson JJ, Mello CC, Kramer JM: Genetic identification, sequence, and alternative splicing of the *Caeanorhabditis elegans* $\alpha2(IV)$ collagen gene. J Cell Biol 1993;123:255–264.

40 Mariyama M, Leinonen A, Mochizuki T, Tryggvason K, Reeders S: Complete primary structure of the human $\alpha3(IV)$ collagen chain: Coexpression of the $\alpha3(IV)$ and $\alpha4(IV)$ collagen chains in human tissues. J Biol Chem 1994;269:23013–23017.

41 Leinonen A, Mariyama M, Mochizuki T, Tryggvason K, Reeders S: Complete primary structure of the human type IV collagen $\alpha4(IV)$ chain: Comparison with structure and expression of the other $\alpha(IV)$ chains. J Biol Chem 1994;269:26172–26177.

42 Pihlajaniemi T, Pohjalainen E-R, Myers J: Complete primary structure of the triple helical region and the carboxyl-terminal domain of a new type IV collagen chain, $\alpha5(IV)$. J Biol Chem 1990;265:13758–13766.

43 Zhou J, Hertz JM, Leinonen A, Tryggvason K: Complete amino acid sequence of the human $\alpha5(IV)$ collagen chain and identification of a single-base mutation in exon 23 converting glycine 521 in the collagenous domain to cysteine in an Alport syndrome patient. J Biol Chem 1992;267:12475–12481.

44 Oohashi T, Sugimoto M, Mattei M-G, Ninomiya Y: Identification of a new collagen IV chain, $\alpha6(IV)$, by cDNA isolation and assignment of the gene to chromosome Xq22, which is the same locus for COL4A5. J Biol Chem 1994;269:7520–7526.

45 Zhou J, Ding M, Zhao Z, Reeders ST: Complete primary structure of the sixth chain of human basement membrane collagen, $\alpha6(IV)$. Isolation of the cDNA for $\alpha6(IV)$ and comparison with five other type IV collagen chains. J Biol Chem 1994;269:13193–13199.

46 Tryggvason K: Mutations in type IV collagen genes and Alport phenotypes; in Tryggvason (ed): Molecular pathology and genetics of Alport syndrome. Contrib Nephrol. Basel, Karger, 1995, vol 117, pp 154–171.

47 Kivirikko KI, Myllylä R: Post-translational processing of procollagens. Ann NY Acad Sci 1985;460:187–201.

48 Timpl R, Wiedemann H, van Delden V, Furthmayr H, Kühn K: A network model for the organization of type IV collagen molecules in basement membranes. Eur J Biochem 1981;120:203–211.

49 Yurchenco PD, Furthmayr H: Self-assembly of basement membrane collagen. Biochemistry 1984;23:1839–1850.

50 Langeweld JPM, Noelken ME, Hård K, Todd P, Vliegenthart JFG, Rouse J, Hudson BG: Bovine glomerular basement membrane: Location and structure of the asparagine-linked oligosaccharide units and their potential role in the assembly of the 7S collagen IV tetramer. J Biol Chem 1991;266:2622–2631.

51 Yurchenco PD, O'Rear J: Supramolecular organization of basement membranes; in Rohrbach DH, Timpl R (eds): Molecular and Cellular Aspects of Basement Membranes. San Diego, Academic Press, 1993, pp 19–47.

52 Vandenberg P, Kern A, Ries A, Luckenbill-Edds L, Mann K, Kühn K: Characterization of a type IV collagen major cell binding site with affinity to the $\alpha1\beta1$ and the $\alpha2\beta1$ integrins. J Cell Biol 1991;113:1475–1483.

53 Kern A, Eble J, Golbik R, Kühn K: Interaction of type IV collagen with the isolated integrins $\alpha1\beta1$ and $\alpha2\beta1$. Eur J Biochem 1993;215:151–159.

54 Ruoslahti E: Integrins. J Clin Invest 1991;87:1–5.

55 Eble JA, Golbik R, Mann K, Kühn K: The $\alpha1\beta1$ integrin recognition site of the basement membrane collagen molecule $[\alpha1(IV)]_2 \alpha2(IV)$. EMBO J 1993;12:4795–4802.

56 Neilson EG, Kalluri R, Sun MJ, Gunwar S, Danoff T, Mariyama M, Myers JC, Reeders ST, Hudson BG: Specificity of Goodpasture autoantibodies for the recombinant noncollagenous domains of human type IV collagen. J Biol Chem 1993;268:8402–8405.

57 Engel J: Structure and function of laminin; in Rohrbach DH, Timpl R (eds): Molecular and Cellular Aspects of Basement Membranes. San Diego, Academic Press, 1993, pp 147–176.

58 Bruch M, Landwehr R, Engel J: Dissection of laminin by cathepsin G into its long-arm and short-arm structures and localization of regions involved in calcium-dependent stabilization and self-association. Eur J Biochem 1989;185:271–279.

59 Burgeson RE, Chiquet M, Deutzmann R, Ekblom P, Engel J, Kleinman H, Martin GR, Meneguzzi G, Paulsson M, Sanes J, Timpl R, Tryggvason K, Yamada Y, Yurchenco PD: A new nomenclature for laminins. Matrix Biol 1994;14:209–211.

60 Pikkarainen T, Eddy R, Fukushima Y, Byers M, Shows T, Pihlajaniemi T, Saraste M, Tryggvason K: Human laminin B1 chain. A multidomain protein with gene (LAMB1) locus in the q22 region of chromosome 7. J Biol Chem 1987;262:10454–10462.

61 Pikkarainen T, Kallunki T, Tryggvason K: Human laminin B2 chain. Comparison of the complete amino acid sequence with the B1 chain reveals variability in sequence homology between different structural domains. J Biol Chem 1988;263:6751–6758.

62 Nissinen M, Vuolteenaho R, Boot-Handford R, Kallunki P, Tryggvason K: Primary structure of the human laminin A chain. Limited expression in human tissues. Biochem J 1991;276:369–379.

63 Sasaki M, Kato S, Kohno K, Martin GR, Yamada Y: Sequence of the cDNA encoding the laminin B1 chain reveals a multidomain protein containing cysteine-rich repeats. Proc Natl Acad Sci USA 1987;84:935–939.

64 Durkin ME, Bartos BB, Liu S-H, Phillips SL, Chung AE: Primary structure of the mouse laminin B2 chain and comparison with laminin B1. Biochemistry 1988;27:5198–5204.

65 Sasaki N, Yamada Y: The laminin B2 chain has a multidomain structure homologous to the B1 chain. J Biol Chem 1987;262:17111–17117.

66 Montell DJ, Goodman CS: *Drosophila* substrate adhesion molecule: Sequence of laminin B1 chain reveals domains of homology with mouse. Cell 1988;53:463–473.

67 Chi H-C, Hui C-F: Primary structure of the *Drosophila* laminin B2 chain and comparison with human, mouse, and *Drosophila* laminin B1 and B2 chains. J Biol Chem 1989;264:1543–1550.

68 Montell DJ, Goodman CS: *Drosophila* laminin: Sequence of B2 subunit and expression of all three subunits during embryogenesis. J Cell Biol 1989;109:2441–2453.

69 Kusche-Gullberg M, Garrison K, McKrell AJ, Fessler LI, Fessler JH: Laminin A chain: Expression during *Drosophila* development and genomic sequence. EMBO J 1992;11:4519–4527.

70 Ehrig K, Leivo I, Argraves WS, Ruoslahti E, Engvall E: Merosin, a tissue-specific basement membrane protein, is a laminin-like protein. Proc Natl Acad Sci USA 1990;87:3264–3268.

71 Vuolteenaho R, Nissinen M, Sainio K, Byers M, Eddy R, Hirvonen H, Shows TB, Sariola H, Engvall E, Tryggvason K: Human laminin M chain (merosin): Complete primary structure, chromosomal assignment, and expression of the M and A chain in human fetal tissues. J Cell Biol 1994;124: 381–394.

72 Marinkovich PM, Lunstrum GP, Burgeson RE: The anchoring filament protein kalinin is synthesized and secreted as a high molecular weight precursor. J Biol Chem 1992;267:17900–17906.

73 Ryan MC, Tizard R, VanDevanter DR, Carter WG: Cloning of the LAMA3 gene encoding the α3 chain of the adhesive ligand epiligrin. J Biol Chem 1994;269:22779–22787.

74 Hunter DD, Shah V, Merlie JP, Sanes JR: A laminin-like adhesive protein concentrated in the synaptic cleft of the neuromuscular junction. Nature 1989;338:229–234.

75 Iivanainen A, Vuolteenaho R, Sainio K, Eddy R, Shows TB, Sariola H, Tryggvason K: The human laminin β2 chain (S-laminin): Structure, expression in fetal tissues and chromosomal assignment of the LAMB2 gene. Matrix Biol 1994;14:489–497.

76 Gerecke DR, Wagman DW, Champliaud MF, Burgeson RE: The complete primary structure for a novel laminin chain, the laminin B1k chain. J Biol Chem 1994;269:11073–11080.

77 O'Rear JJ: A novel laminin B1 chain variant in avian eye. J Biol Chem 1992;267:20555–20557.

78 Kallunki P, Sainio K, Eddy R, Byers M, Kallunki T, Sariola H, Beck K, Hirvonen H, Shows TB, Tryggvason K: A truncated laminin chain homologous to the B2 chain: Structure, spatial expression, and chromosomal assignment. J Cell Biol 1992;119:679–693.

79 Sugiyama S, Utani S, Kozak CA, Yamada Y: Cloning and expression of the mouse laminin γ2 (B2t) chain, a subunit of epithelial cell laminin. Eur J Biochem 1995;228:120–128.

80 Rousselle P, Lunstrum GP, Keene DR, Burgeson RE: Kalinin: An epithelium-specific basement membrane adhesion molecule that is a component of anchoring filaments. J Cell Biol 1991;114: 567–576.

81 Carter WG, Ryan NC, Gahr PJ: Epiligrin, a new cell adhesion ligand for integrin α3β1 in epithelial basement membranes. Cell 1991;65:599–610.

82   Marinkovich MP, Lunstrum GP, Keene DR, Burgeson RE: The dermal-epidermal junction of human skin contains a novel laminin variant. J Cell Biol 1992;119:695–703.

83   Engvall E, Earwicker D, Haaparanta T, Ruoslahti E, Sanes JR: Distribution and isolation of four laminin variants; tissue restricted distribution of heterotrimers assembled from five different subunits. Cell Regul 1990;1:731–740.

84   Green TL, Hunter DD, Chan W, Merlie JP, Sanes JR: Synthesis and assembly of the synaptic cleft protein S-laminin by cultured cells. J Biol Chem 1992;267:2014–2022.

85   Boot-Handford RP, Kurkinen M, Prockop DJ: Steady-state levels of mRNAs coding for the type IV collagen and laminin polypeptide chains of basement membranes exhibit marked tissue-specific stoichiometric variations in the rat. J Biol Chem 1987;262:12474–12478.

86   Kleinman HK, Ebihara I, Killen PD, Sasaki M, Cannon FB, Yamada Y, Martin GR: Genes for basement membrane proteins are coordinately expressed in differentiating F9 cells but not in normal adult murine tissues. Dev Biol 1987;122:373–378.

87   Klein G, Ekblom M, Fecker L, Timpl R, Ekblom P: Differential expression of laminin A and B chains during development of embryonic mouse organs. Development 1990;110:823–837.

88   Sanes JR, Engvall E, Butkowski R, Hunter DD: Molecular heterogeneity of basal laminae: Isoforms of laminin and collagen IV at the neuromuscular junction and elsewhere. J Cell Biol 1990;111: 1685–1699.

89   Hsiao LL, Engvall E, Peltonen J, Uitto J: Expression of laminin isoforms by peripheral nerve-derived connective tissue cells in culture. Comparison with epitope distribution in normal human nerve and neural tumors in vivo. Lab Invest 1993;68:100–108.

90   Ekblom M, Klein G, Mugrauer G, Fecker L, Deutzmann R, Timpl R, Ekblom P: Transient and locally restricted expression of laminin A chain mRNA by developing epithelial cells during kidney organogenesis. Cell 1990;60:337–346.

91   Laurie GW, Horikoshi S, Killen PD, Segui-Real B, Yamada Y: In situ hybridization reveals temporal and spatial changes in cellular expression of mRNA for laminin receptor, laminin, and basement membrane (type IV) collagen in the developing kidney. J Cell Biol 1989;109:1351–1362.

92   Thomas T, Dziadek M: Genes coding for basement membrane glycoproteins laminin, nidogen, and collagen IV are differentially expressed in the nervous system and by epithelial, endothelial, and mesenchymal cells of the mouse embryo. Exp Cell Res 1993;208:54–67.

93   Paulsson M, Deutzmann R, Timpl R, Dalzoppo D, Odermatt E, Engel J: Evidence for coiled-coil α-helical regions in the long arm of laminin. EMBO J 1985;4:309–316.

94   Hunter I, Schulthess T, Engel J: Laminin chain assembly by triple and double stranded coiled-coil structures. J Biol Chem 1992;267:6006–6011.

95   Arumugham RG, Hsieh TC-Y, Tanzer ML, Laine RA: Structures of the asparagine-linked sugar chains of laminin. Biochim Biophys Acta 1986;883:112–126.

96   Fujiwara S, Shinkai H, Deutzmann R, Paulsson M, Timpl R: Structure and distribution of N-linked oligosaccharide chains on various domains of mouse tumour laminin. Biochem J 1988;252: 453–461.

97   Knibbs RN, Perini F, Goldstein IJ: Structure of the major concanavalin A reactive oligosaccharides of the extracellular matrix component laminin. Biochemistry 1989;28:6379–6392.

98   Chandrasekaran S, Dean JW III, Giniger MS, Tanzer ML: Laminin carbohydrates are implicated in cell signalling. J Cell Biochem 1991;46:115–124.

99   Wu C, Friedman R, Chung AE: Analysis of the assembly of laminin and the laminin-entactin complex with laminin chain-specific monoclonal and polyclonal antibodies. Biochemistry 1988;27: 8780–8787.

100 Tanzer ML, Chandrasekaran S, Dean JW III, Giniger MS: Role of laminin carbohydrates on cellular interaction. Kidney Int 1993;43:66–72.

101 Peters BP, Hartle RJ, Krzesicki RF, Kroll TG, Perini F, Balun JE, Goldstein IJ, Ruddon RW: The biosynthesis, processing, and secretion of laminin by human choriocarcinoma cells. J Biol Chem 1985;260:14732–14742.

102 Tokida Y, Aratani Y, Morita A, Kitagawa Y: Production of two variant laminin forms by endothelial cells and shift of their relative levels by angiostatic steroids. J Biol Chem 1990;265:18123–18129.

103 Ekblom P: Basement membranes in development; in Rohrbach DH, Timpl R (eds): Molecular and Cellular Aspects of Basement Membranes. San Diego, Academic Press, 1993, pp 359–383.

104 Yamada KM: Adhesive recognition sequences. J Biol Chem 1991;266:12809–12812.

105 Mecham RP: Receptors for laminin on mammalian cells. FASEB J 1991;5:2538–2546.

106 Delwel GO, Hogervorst F, Kuikman I, Paulsson M, Timpl R, Sonnenberg A: Expression and function of the cytoplasmic variants of the integrin α6 subunit in transfected K562 cells. Activation-dependent adhesion and interaction with isoforms of laminin. J Biol Chem 1993;268:25865–25875.

107 Gehlsen KR, Dickerson K, Argraves WS, Engvall E, Ruoslahti E: Subunit structure of a laminin-binding integrin and localization of its binding site on laminin. J Biol Chem 1989;264:19034–19038.

108 Sonnenberg A, Linders CJT, Modderman PW, Damsky CH, Aumailley M, Timpl R: Integrin recognition of different cell-binding fragments of laminin (P1, E3, E8) and evidence that α6β1 but not α6β4 functions as a major receptor for fragment E8. J Cell Biol 1990;110:2145–2155.

109 Auth D, Brawerman G: A 33-kDa polypeptide with homology to the laminin receptor: Component of translation machinery. Proc Natl Acad Sci USA 1992;89:4368–4372.

110 Pulkkinen L, Christiano AM, Airenne T, Haakana H, Tryggvason K, Uitto J: Mutations in the γ2 chain gene (LAMC2) of kalinin/laminin 5 in the junctional forms of epidermolysis bullosa. Nat Genet 1994;6:293–297.

111 Timpl R, Dziadek M, Fujiwara S, Nowack H, Wick G: Nidogen: A new, self-aggregating basement membrane protein. Eur J Biochem 1983;137:455–465.

112 Carlin B, Jaffe R, Bender B, Chung AE: Entactin, a novel basal lamina associated sulfated glycoprotein. J Biol Chem 1981;256:5209–5214.

113 Desjardins M, Bendayan M: Ontogenesis of glomerular basement membrane: Structural and functional properties. J Cell Biol 1991;113:689–700.

114 Paulsson M, Dziadek M, Suchanek C, Huttner WB, Timpl R: Nature of sulphated macromolecules in mouse Reichert's membrane: Evidence for tyrosine-O-sulfate in basement-membrane proteins. Biochem J 1985;231:571–579.

115 Paulsson M, Deutzmann R, Dziadek M, Nowack H, Timpl R, Weber S, Engel J: Purification and structural characterization of intact and fragmented nidogen obtained from a tumor basement membrane. Eur J Biochem 1986;156:467–478.

116 Nagayoshi T, Sanborn D, Hickok NJ, Olsen DR, Fazio MJ, Chu M-L, Knowlton R, Mann K, Deutzmann R, Timpl R, Uitto J: Human nidogen: Complete amino acid sequence and structural domains deduced from cDNAs, and evidence for polymorphism of the gene. DNA 1989;8:581–594.

117 Mann K, Deutzmann R, Aumailley M, Timpl R, Raimondi L, Yamada Y, Pan T, Conaway D, Chu M-L: Amino acid sequence of mouse nidogen, a multidomain basement membrane protein with binding activity for laminin, collagen IV and cells. EMBO J 1989;8:65–72.

118 Durkin ME, Chakravarti S, Bartos BB, Liu S-H, Friedman RL, Chung AE: Amino acid sequence and domain structure of entactin. Homology with epidermal growth factor precursor and low density lipoprotein receptor. J Cell Biol 1988;107:2749–2756.

119 Fox JW, Mayer U, Nischt R, Aumailley M, Reinhardt D, Wiedemann H, Mann K, Timpl R, Krieg T, Engel J, Chu M-L: Recombinant nidogen consists of three globular domains and mediates binding of laminin to collagen IV. EMBO J 1991;10:3137–3146.

120 Mann K, Deutzmann R, Timpl R: Characterization of proteolytic fragments of the laminin-nidogen complex and their activity in ligand-binding assays. Eur J Biochem 1988;178:71–80.

121 Gerl M, Mann K, Aumailley M, Timpl R: Localization of a major nidogen-binding site to domain III of laminin B2 chain. Eur J Biochem 1991;202:167–174.

122 Mayer U, Nischt R, Pöschl E, Mann K, Fukuda K, Gerl M, Yamada Y, Timpl R: A single EGF-like motif of laminin is responsible for high affinity nidogen binding. EMBO J 1993;12:1879–1885.

123 Aumailley M, Wiedemann H, Mann K, Timpl R: Binding of nidogen and the laminin-nidogen complex to basement membrane collagen type IV. Eur J Biochem 1989;184:241–248.

124 Battaglia C, Mayer U, Aumailley M, Timpl R: Basement-membrane heparan sulfate proteoglycan binds to laminin by its heparan sulfate chains and to nidogen by sites in the protein core. Eur J Biochem 1992;208:359–366.

125 Chung AE, Dong L-J, Wu C, Durkin ME: Biological functions of entactin. Kidney Int 1993;43: 13–19.

126 Timpl R, Aumailley M: Other basement membrane proteins and their calcium-binding potential; in Rohrbach DH , Timpl R (eds): Molecular and Cellular Aspects of Basement Membranes. San Diego, Academic Press, 1993, pp 211–235.

127 Kjellen L, Lindahl U: Proteoglycans: Structures and interactions. Annu Rev Biochem 1991;60: 443–475.

128 Noonan DM, Hassell JR: Proteoglycans of basement membranes; in Rohrbach DH, Timpl R (eds): Molecular and Cellular Aspects of Basement Membranes. San Diego, Academic Press, 1993, pp 189–210.

129 Kanwar YS, Farquhar MG: Anionic sites in the glomerular basement membrane: In vivo and in vitro localization to the laminae rarae by cationic probes. J Cell Biol 1979;81:137–153.

130 Kanwar YS, Farquhar MG: Presence of heparan sulfate in the glomerular basement membrane. Proc Natl Acad Sci USA 1979;76:1303–1307.

131 Kanwar YS, Linker A, Farquhar MG: Increased permeability of the glomerular basement membrane to ferritin after removal of glycosaminoglycans (heparan sulfate) by enzyme digestion. J Cell Biol 1980;86:688–693.

132 Hassell JR, Robey P, Barrach H, Wilczek J, Rennard S, Martin GR: Isolation of a heparran sulfate-containing proteoglycan from basement membrane. Proc Natl Acad Sci USA 1980;77:4494–4498.

133 Iozzo RV: Biosynthesis of heparan sulfate proteoglycan by human colon carcinoma cells and its localization at the cell surface. J Cell Biol 1984;99:403–417.

134 Saku T, Furthmayr H: Characterization of the major heparan sulfate proteoglycan secreted by bovine aortic endothelial cells in culture. Homology to the large molecular weight molecule of basement membranes. J Biol Chem 1989;264:3514–3523.

135 Heremans A, van der Schueren B, de Cock B, Paulsson M, Cassiman J-J, van den Berghe H, David G: Matrix-associated heparan sulfate proteoglycan: Core protein-specific monoclonal antibodies decorate the pericellular matrix of connective tissue cells and the stromal side of basement membranes. J Cell Biol 1989;109:3199–3211.

136 Mohan PS, Spiro RG: Characterization of heparan sulfate proteoglycan from calf lens capsule and proteoglycans synthesized by cultured lens epithelial cells. Comparison with other basement membrane proteoglycans. J Biol Chem 1991;266:8567–8575.

137 Klein DJ, Brown DM, Oegema TR, Brenchley PE, Andersson JC, Dickinson MAJ, Horigan EA, Hassell JR: Glomerular basement membrane proteoglycans are derived from a large precursor. J Cell Biol 1988;106:963–970.

138 Paulsson M, Yurchenco PD, Ruben GC, Engel J, Timpl R: Structure of low density heparan sulfate proteoglycan isolated from a mouse tumor basement membrane. J Mol Biol 1987;197:297–313.

139 Noonan DM, Fulle A, Valente P, Cai S, Horigan E, Sasaki N, Yamada Y, Hassell JR: The complete sequence of perlecan, a basement membrane heparan sulfate proteoglycan, reveals extensive similarity with laminin A chain, low density lipoprotein receptor, and the neural cell adhesion molecule. J Biol Chem 1991;266:22939–22947.

140 Kallunki P, Tryggvason K: Human basement membrane heparan sulfate proteoglycan core protein: A 467-kD protein containing multiple domains resembling elements of the low density lipoprotein receptor, laminin, neural cell adhesion molecules, and epidermal growth factor. J Cell Biol 1992; 116:559–571.

141 Noonan DM, Hassell JR: Perlecan, the large low-density proteoglycan of basement membranes: Structure and variant forms. Kidney Int 1993;43:53–60.

142 Vlodavsky I, Bar-Shavit R, Korner G, Fuks Z: Extracellular matrix-bound growth factors, enzymes, and plasma proteins; in Rohrbach DH, Timpl R (eds.): Molecular and Cellular Aspects of Basement Membranes. San Diego, Academic Press, 1993, pp 327–343.

143 Grant DS, Leblond CP, Kleinman HK, Inoue S, Hassell JR: The incubation of laminin, collagen IV, and heparan sulfate proteoglycan at 35 °C yields basement membrane-like structures. J Cell Biol 1989;108:1567–1574.

144 Fujiwara S, Wiedemann H, Timpl R, Lustig A, Engel J: Structure and interactions of heparan sulfate proteoglycans from a mouse tumor basement membrane. Eur J Biochem 1984;143:145–157.

145 Kanwar YS, Veis A, Kimura JH, Jakubowski ML: Characterization of heparan sulfate proteoglycan of glomerular basement membranes. Proc Natl Acad Sci USA 1984;81:762–766.

146 Ledbetter SR, Tyree B, Hassell JR, Horigan EA: Identification of the precursor protein to basement membrane heparan sulfate proteoglycans. J Biol Chem 1985;260:8106–8113.

147 Kato M, Koike Y, Ito Y, Suzuki S, Kimata K: Multiple forms of heparan sulfate proteoglycans in the Engelbreth-Holm-Swarm mouse tumour: The occurrence of high density forms bearing both heparan and chondroitin sulfate side chains. J Biol Chem 1987;262:7180–7188.

148 McCarthy KJ, Accavitti MA, Couchman JR: Immunological characterization of basement membrane-specific chondroitin sulfate proteoglycan. J Cell Biol 1989;109:3187–3198.

149 Quigg RJ: Isolation of a novel complement regulatory factor (GCRF) from glomerular epithelial cells. Kidney Int 1991;40:668–676.

150 Beavan LA, Davies M, Mason RM: Renal glomerular proteoglycans. An investigation of their synthesis in vivo using a technique for fixation in situ. Biochem J 1988;251:411–418.

151 Dziadek M, Paulsson M, Aumailley N, Timpl R: Purification and tissue distribution of a small protein (BM-40) extracted from a basement membrane tumour. Eur J Biochem 1986;161:455–464.

152 Termine JD, Kleinman HK, Whitson SW, Conn KM, McGarvey ML, Martin GR: Osteonectin, a bone-specific protein linking mineral to collagen. Cell 1981;26:99–105.

153 Sage H, Johnson C, Bornstein P: Characterization of a novel serum albumin-binding glycoprotein secreted by endothelial cells in culture. J Biol Chem 1984;259:3993–4007.

154 Robinson LK, Murrah VA, Mayer MP, Rohrbach DH: Characterization of a novel glycoprotein isolated from the basement membrane matrix of the Engelbreth-Holm-Swarm tumor. J Biol Chem 1989;264:5141–5147.

155 Abrahamson DR: Structure and development of the glomerular capillary wall and basement membrane. Am J Physiol 1987;253:F783–F794.

156 Thomas T, Dziadek M: Genes coding for basement membrane glycoproteins laminin, nidogen, and collagen IV are differentially expressed in the nervous system and by epithelial, endothelial, and mesenchymal cells of the mouse embryo. Exp Cell Res 1993;208:54–67.

157 Couchman JR, Abrahamson DR, McCarthy KJ: Basement membrane proteoglycans and development. Kidney Int 1993;43:79–84.

158 Ardaillou N, Bellon G, Nivetz M-P, Rakotoarison S, Ardaillou R: Quantification of collagen synthesis by cultured human glomerular cells. Biochim Biophys Acta 1989;991:445–452.

159 Aumailley M, Battaglia C, Mayer U, Reinhardt D, Nischt R, Timpl R, Fox JW: Nidogen mediates the formation of ternary complexes of basement membrane components. Kidney Int 1993;443:7–12.

160 Haskin CL, Cameron IL, Rohrbach DH: Role of water of hydration in filtration function of proteoglycans of basement membranes; in Rohrbach DH, Timpl R (eds): Molecular and Cellular Aspects of Basement Membranes. San Diego, Academic Press, 1993, pp 107–117.

Taina Pihlajaniemi, MD, Collagen Research Unit, Biocenter, and Department of Medical Biochemistry, University of Oulu, Kasaanintie 52A, FIN–90220 Oulu (Finland)

Tryggvason K (ed): Molecular Pathology and Genetics of Alport Syndrome.
Contrib Nephrol. Basel, Karger, 1996, vol 117, pp 80–104

The α Chains of Type IV Collagen

Jing Zhou, Stephen T. Reeders

Renal Division, Department of Medicine, Brigham and Women's Hospital,
Harvard Medical School, Boston, Mass., USA

Introduction

BMs contain six genetically distinct type IV collagen α-chains, each of
which is encoded by its own gene. The six α chains, α1(IV)–α6(IV) are encoded
by the *COL4A1-6* genes in man and the *col4a1-6* genes in the mouse. The
COL4A1-6 genes are arranged in three pairs, *COL4A1* and *COL4A2, COL4A3*
and *COL4A4,* and *COL4A5* and *COL4A6,* which reside on human chromosomes 13, 2q35, and Xq22, respectively [1–3]. The six chains are closely related
in structure, and form a distinct family of collagens. The primary structures
of all six chains have been determined in man and several of them have also
been determined in mouse. Table 1 summarizes the properties of the six chains
and their genes.

Despite the rapid explosion of our knowledge of the type IV collagens
and the recent evidence that they are involved in at least two distinct disease
states, relatively little is known about the discriminant functions of the six
chains. Initial inferences can, however, be made by careful comparison of the
primary structures and tissue distributions of the chains and analysis of the
evolutionary relationships of their genes. In this chapter we review the similarities and differences between the chains in an attempt to lay the groundwork
for experimental approaches to elucidation of their possible functions.

The α chains can be classified in several different ways. The α1(IV) and
α2(IV) chains, which were the first to be described, are abundant and have been
termed 'classical' chains to distinguish them from the four 'novel' chains which
are less abundant [4]. Indeed the α5(IV) and α6(IV) chains were discovered
through the cloning of their genes rather than through biochemical analysis of
BMs. The classical chains appear to be ubiquitous, existing in most if not in all
BMs. By contrast, the α3(IV)–α6(IV) chains have restricted tissue distributions.

Table 1. Properties of six type IV collagen α chains and their genes

α chain	Gene symbol	Chromosome	Exon numbers	Family	Length of mature chains (residues)	Length of NC1 domains (residues)	Reference
Human α1(IV)	COL4A1	13q34	52	α1-like	1,642	229	1, 6, 32, 86, 87
Mouse α1(IV)	–	–	–	α1-like	1,642	229	27, 28
Human α2(IV)	COL4A2	13	–	α2-like	1,676	227	1, 6, 29, 31, 95
Mouse α2(IV)	–	–	–	α2-like	1,689	226	30, 88, 91, 92
Human α3(IV)	COL4A3	2q35	–	α1-like	1,642	232	2, 41, 25
Mouse α3(IV)	–	–	–	α1-like	–	233	74
Bovine α3(IV)	–	–	–	α1-like	–	233	40
Human α4(IV)	COL4A4	2q35	–	α2-like	1,652	231	2, 18
Mouse α4(IV)	–	–	–	α2-like	–	231	74
Bovine α4(IV)	–	–	–	α2-like	–	231	42
Human α5(IV)	COL4A5	Xq22	51	α1-like	1,659	229	43, 44, 95
Mouse α5(IV)	–	–	–	α1-like	–	229	74
Human α6(IV)	COL4A6	Xq22	–	α2-like	1,670	228	3, 5, 45
A. summ α2(IV)	–	–	>14	α2-like	1,737	234	11
C. elegans clb2	emb9	III	12	α1-like	1,727	229	8, 9
C. elegans clb1	let2	X	20	α2-like	1,732	231	10
Sea urchin 3α	COLP3α	–	–	α1-like	1,724	226	12
Sea urchin 4α	COLP4α	–	–	α2-like	1,716	225	13
D. melanogaster α(IV)	COL4	25C	9	α1-like	1,752	231	89, 93, 94

The Evolution of the Type IV Collagens

Detailed analysis of the sequences of the α chains has enabled us to determine the probable evolutionary development of the human α chains with considerable confidence. The data predict that a series of duplications of intact ancestral α-chain genes lead to the creation of the six genes found in man. This model of evolution obtains strong support from the genomic organization of the α-chain genes in man and mouse (fig. 1) [5].

Three duplication events occurred during the evolution of the type IV genes. First, an ancestral type IV gene was duplicated so that one copy came

Abbreviations

AS	Alport syndrome	GBM	glomerular basement membrane
ASDL	AS-associated diffuse leiomyomatosis	kb	kilobase(s)
bp	base pair(s)	NC	noncollagenous
BM	basement membrane	PCR	polymerase chain reaction

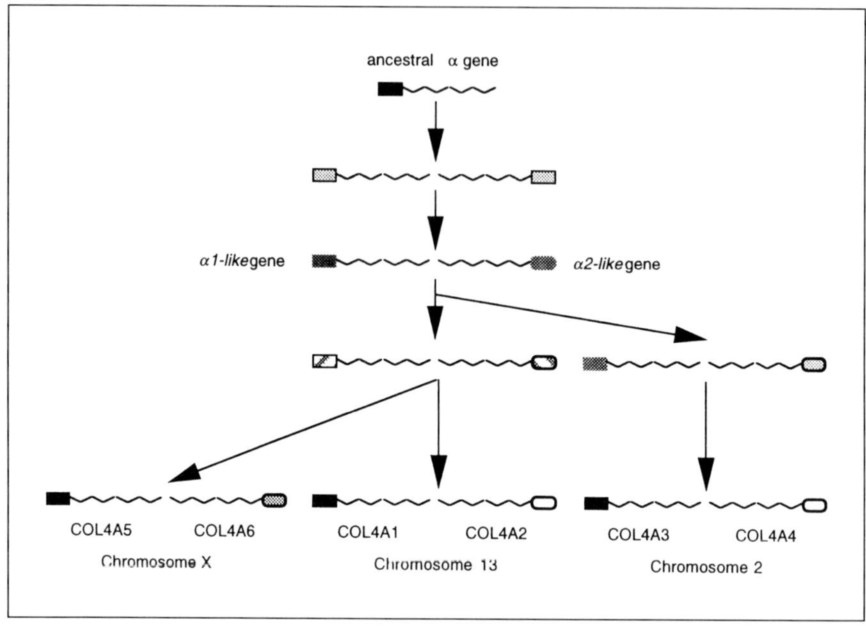

Fig. 1. Evolution of the type IV collagens. The pattern of evolution was deduced from sequence analysis of the NC1 domains of the six known type IV collagen chains. An ancestral type IV gene was duplicated so that one copy came to lie head-to-head with the other. These genes then diverged to give the precursors of the current *α1-like* and *α2-like* genes. The paired complex underwent a second round of duplication resulting in two new complexes. One of these subsequently evolved to become the current COL4A3 and COL4A4. The other complex evolved and, at some later date, underwent a further round of duplication giving rise to four genes in all.

to lie head-to-head with the other. These genes then diverged to give two quite distinct type IV collagen genes, the precursors of the current *α1-like* and *α2-like* genes. Several differences between the precursors remain as features which distinguish the daughter genes to this day. These conserved differences presumably represent important functional features of the *α1-like* and *α2-like* classes. The complex of two head-to-head genes underwent a second round of duplication resulting in two new complexes. One of these subsequently evolved to become the current α3(IV) and α4(IV) genes which still lie head-to-head on chromosome 2 in man [2]. The other complex evolved and, at some later data, underwent a further round of duplication giving rise to four genes in all. These are the current α1(IV) and α2(IV) genes, which lie head-to-head on chromosome 13 in man [1, 6], and the α5(IV) and α6(IV) genes, which lie head-to-head on the X chromosome in man [3]. The α1(IV), α3(IV) and α5(IV)

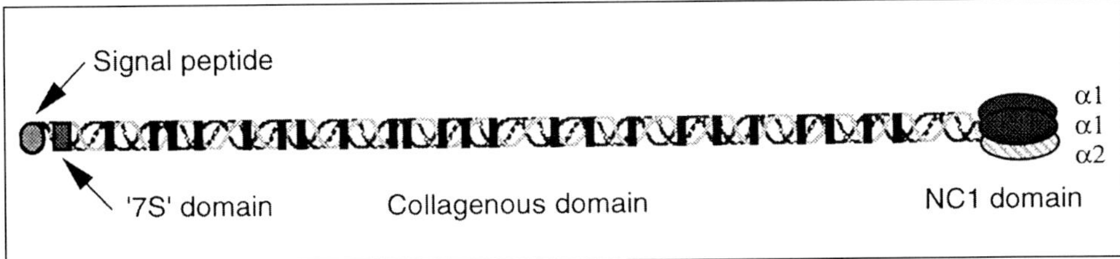

Fig. 2. Diagram of type IV collagen monomer with typical chain composition of two α1(IV) chains and one α2(IV) chain. The individual α chains are distinguished by different gray levels. The main triple helix, interrupted by noncollagenous sequences (indicated by vertical bars), terminates in a globular domain composed of three NC1 domains (shown as oval). The signal peptides and the short amino-terminal noncollagenous domains (7S domain) are indicated by arrows.

chains form the *α1-like* class whereas the α2(IV), α4(IV) and α6(IV) form the *α2-like* class. It should be noted that only a single gene has been found in *Drosophila* [7]. There are α1-like and α2-like chains in nematodes, including *Caenorhabditis elegans and Ascaris suum* [8–11] as well as in sea urchins [12, 13]. These chains have distinct resemblance to the mammalian *α1-like* and *α2-like* chains but they are located on different chromosomes [8]. Presumably, the α1/α2 divergence occurred before the split between the roundworm and the phyla that have subsequently given rise to mammals. The substantial interval that has elapsed since this species divergence enables us to make several inferences about the likely functional importance of the common structural features that are found in either class in the roundworm and mammal (see below).

The Primary Structures of the Type IV Collagen Chains

In this section we will deal with the homologous structural features that are common to all type IV collagens. Since these features are highly conserved, being present in all type IV molecules from the roundworm to man, we can assume that they confer important functional properties that define and distinguish the type IV collagens.

Each type IV collagen molecule (monomer) is composed of three α chains. Each α chain consists of three distinct structural domains: an amino-terminal cysteine-rich domain of 14–23 amino acids (7S), a collagenous domain of ~1,400 amino acids, and a ~230 amino acid globular carboxyl-terminal (NC1) domain (fig. 2). A characteristic feature of the collagenous domain is the presence of between 21 and 26 interruptions in the Gly-X-Y triplet sequences. The type IV chains have signal peptides of between 21 and 38 amino acids followed by classical signal peptide cleavage sites. The predicted cleavages create mature molecules of between 1,642 and 1,670 amino acids in man.

The C-Terminal NC1 Domains

The C-terminal collagenase-resistant domains have been studied extensively at the biochemical level since they can be fractionated on gels following collagenase treatment of BMs.

Study of NC1 interactions provides insight into the stoichiometry of the chains in triple helices. Since EHS tumor and the kidney, particularly the glomerulus, are the predominant sources of type IV collagen for biochemical analysis, much of our knowledge of the stoichiometries and properties of the NC1 domains reflects these tissues. The NC1 domains are present in BMs as hexamers which can be dissociated into dimers and monomers (fig. 3). The dimers are stabilized by disulfide bonds. In all, the 24 cysteine residues in each pair of NC1 domains take part in 12 disulfide bonds, 8 of which are intermolecular. Siebold et al. [57] have determined the structure of these bonds in the $\alpha 1(IV)$ chain (fig. 3). The striking degree of conservation between the positions of the 12 cysteine residues in all of the known NC1 domains suggests that the cysteine pairing is the same for all NC1 dimers.

The six α chains in man allow for a total of 21 possible NC1 dimers. The current data, obtained by two different approaches, suggest that homodimerization is the predominant form for the $\alpha 1$-$\alpha 4(IV)$ chains [14–16]. Johansson and Gunwar also found smaller amounts of $\alpha 1/\alpha 3(IV)NC1$ heterodimers while Johansson found $\alpha 2/\alpha 4(IV)NC1$ heterodimers. Little is known about the dimer stoichiometries of the $\alpha 5(IV)$ and $\alpha 6(IV)$ chains but evidence of $\alpha 3/\alpha 5(IV)NC1$ heterodimer has been provided by Kleppel et al. [15] and Derry et al. [16]. Derry has also found $\alpha 1/\alpha 5(IV)NC1$ heterodimers.

The stoichiometries of the triple helices are less well known. The six chains allow for a total of 56 different triple helices. In EHS tumors the $\alpha 1(IV)_2 \cdot \alpha 2(IV)$ heterotrimer is the predominant form. In the GBM, at least two type IV collagen molecules exist, $\alpha 1(IV)_2 \cdot \alpha 2(IV)$ and $\alpha 3(IV)_2 \cdot \alpha 4(IV)$ heterotrimers [14]. Homotrimers composed of $\alpha 1(IV)_3$ and $\alpha 3(IV)_3$ have also been reported [17]. Very little is known about the stoichiometries of triple helices present at other sites.

Comparison of the NC1 Domains

Inspection of the NC1 domain reveals that it arose as a result of the tandem duplication of a hemi-NC1 domain of about 115 amino acids. This duplication clearly antedated the formation of the $\alpha 1$-like and $\alpha 2$-like ancestors since all known type IV collagens have the same NC1 structure. Residues and motifs that are conserved between both halves are likely to be important in

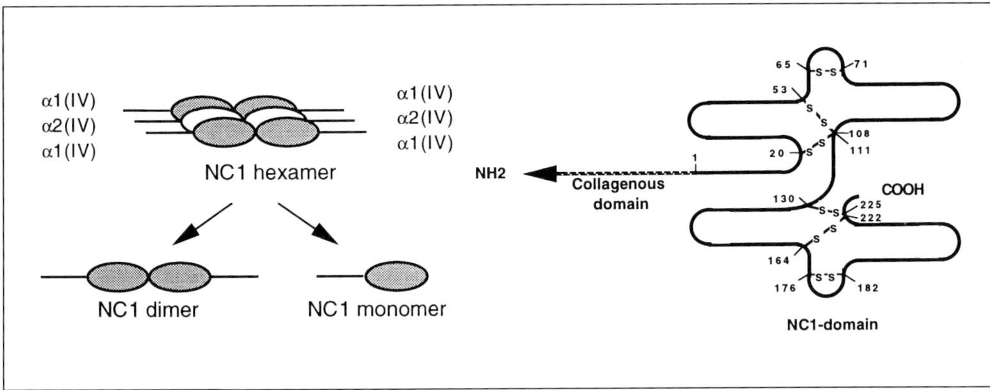

Fig. 3. Illustration of NC1 domain structure and associations. *A* A hexamer isolated from native tissues after collagenase digestion is composed of NC1 domains of six α chains. Hexamers can disassociate into dimers and monomers under denaturing conditions. *B* Schematic illustration of the two subdomains and the intra-chain disulfide bonds in the NC1 domain of a single α1(IV) chain. The numbering of the residues starts from the amino-terminal end of the NC1 domain. Modified from Siebold et al. [57].

the secondary structures of the NC1 domains. For example, each hemi-NC1 contains 6 cysteine residues. Furthermore, as discussed above, the two halves have identical disulfide-bonding arrangements. In each half, the relative positions of the cysteines are also conserved. Analysis of the intact NC1 domains reveals that there are segments of conservation and other segments in which there is considerable variation from chain to chain [5, 18]. For example, in the human chains, the 20 residues N-terminal to the penultimate cysteine show very little homology. It is striking that the corresponding set of residues in the first hemi-NC1 domains also differ substantially. By contrast, the residues immediately preceding these variable stretches are highly conserved in both hemi-NC1 domains. The variable regions are possible sites for selective binding of like chains and may mediate NC1 homodimerization.

We have concluded that the type IV collagens can be divided into two classes (see above). Inspection of the NC1 domains reveals several structural features that distinguish the two classes. These features are prime candidates for functionally important properties that differentiate the two classes. For example, at position 1605 (α6(IV) numbering), the *α1-like* chains all have a tyrosine residue whereas the α2-like chains have arginine or lysine residues [5]. More detailed evaluation reveals that there are longer stretches that are common to either the *α1-like* or *α2-like* classes but not both.

Despite these observations, we caution that not all features that are conserved during evolution are necessarily relevant. Chance also plays a large part in the sequences we observe today. Nevertheless, the collagenous domain provides an important calibration for the conservation of the NC1 domains. Since the collagenous and NC1 domains of different molecules diverged at the same time, the opportunity for mutation is the same in both domains. Comparison of the collagenous domains reveals that the second and third positions of the Gly-X-Y triplets are barely conserved, whereas the NC1 domains are highly conserved.

The Collagenous Domains and Interruptions

In man the collagenous domains of the six chains are between 1,410 residues (α3(IV) chain) and 1,449 residues (α2(IV) chain). If it is assumed that the interruptions may loop out to varying extents, it is possible to align the collagenous regions of all α chains.

Several lines of evidence suggest that these interruptions, which are a characteristic feature of the type IV collagens, are not simply random insertional mutations that have been tolerated within the collagenous domain during evolution. The most important support for a functional role for these interruptions is the degree of conservation of the lengths and positions of the interruptions within the molecules. Furthermore, there are striking homologies between the corresponding interruptions in different chains. For example, interruption XII consists of two residues with a glycine in the first position in all six chains.

A popular hypothesis to account for the interruptions is that the type IV collagen network requires not only the strength to support the BM network but also the flexibility to buffer the pressure it receives. Some of the interruptions may serve as cell-binding sites [19]. The presence of cysteines in the interruptions also suggest that the interruptions may be sites of inter-chain cross-linking.

Another feature of the collagenous domain is the presence of a short stretch of non-collagenous, cysteine-rich sequence at the amino terminus. This so-called '7S' domain was discovered by its resistance to bacterial collagenase. The four cysteine residues in this domain are present in all six chains and are essential for the cross-linking of four triple helical molecules through disulfide bridges [20–22].

The type IV collagen α-chains are extensively glycosylated [23, 24]. There are $\sim$ 50 hydroxylysine-linked disaccharide sites along the collagenous domain. All the chains except α3(IV) possess an asparagine-like glycosylation site, N-G-T(S), in the first triple helical segment [25].

In the following section, we will discuss the structures of type IV collagen isoforms by dividing them into three groups: the α1(IV) and α2(IV) chains, the α3(IV) and α4(IV) chains, and the α5(IV) and α6(IV) chains.

Table 2. Major features of the six human collagen IV α chains

	α1	α2	α3	α4	α5	α6
Complete translation product (residues)	1,669	1,712	1,670	1,690	1,685	1,691
Mature α chain (residues)	1,642	1,676	1,652	1,652	1,659	1,670
Signal peptide (residues)	27	36	28	38	26	21
Collagenous domain (residues)	1,413	1,449	1,410	1,421	1,430	1,417
NC1 domain (residues)	229	227	232	231	229	228
Number of interruptions in collagenous domain	21	23	23	26	22	25
References	86, 87	29, 31	25	18	44	5, 45

The α1 (IV) and α2 (IV) Chains

These two chains are the most abundant type IV collagens and are therefore the most thoroughly studied [26]. The human α1(IV) and α2(IV) chains migrate as 185 and 170 kD proteins although their calculated molecular weights are 157 and 167 kD, respectively. They appear to exist primarily in a 2:1 ratio. The complete primary structures of the murine homologues of these two chains are known in mouse [27–29].

The major structural features of all human α(IV) chains are summarized in table 2. As compared with the α1(IV) chain, the α2(IV) has one more cysteine in the collagenous domain. With the exception of the two cysteine residues in interruption XIII, which form an intra-chain disulfide bridge that causes part of the collagenous domain to loop out [29] and one cysteine in interruption VIII of the α1(IV) chain, all the cysteines in the two chains are conserved. Almost all the cysteines were conserved between species although the cysteines in interruption XIII are not present in the sea urchin [12]. The α1(IV) and the α2(IV) chains have 63% nucleotide sequence similarity in the NC1 domain but only 26% in the collagenous domain if the obligatory glycine is not used for the analysis [30, 31]. The sequence identity of the bovine, murine and human α1(IV) chains is about 85% in the collagenous domain [28, 32], and 90.6% between mouse and man [28]. The overall sequence identity for α2(IV) is 83.5% between mouse and man. However, the NC domain is highly conserved with 97% sequences identity between the human and murine α1(IV) chains [27]. The invertebrate α1 and α2(IV) chains process all the type IV collagen structure features and have ∼44% overall sequences identity with the human chains. In the 7S and the NC1 domains, however, the sequences homology is ∼52% and ∼60% [13]. Most of the helix interruptions are located at the same position in the α1(IV) and α2(IV) chains. As in the case of

Drosophila, 11 of the 22 imperfections were found in the same position, which might be important for the lateral association of the molecules [33].

The α3(IV) and α4(IV) Chains

The existence of minor type IV collagen subunits, the α3(IV) and α4(IV) chains, has been implicated for several years based on chemical analyses of the GBM. Following digestion of the GBM with bacterial collagenase, proteins containing type IV collagen NC domain-like components distinct from those of the α1 and α2(IV) chains were identified [34, 35]. The different monomeric subunits resulting from collagenous digestion of human GBM have been termed M24, M26, M28 + + +, and M28 +, while the equivalent subunits of bovine BMs have been termed M1a, M1b, M2*, and M3 [36, 37]. M24 (or M1a) and M26 (or M1b) are the NC domains of the α1(IV) and α2(IV) chains, respectively, and M28 + + + (or M2*) and M28 + (or M3) are the NC domains of two novel collagen chains, designated α3 and α4(IV), respectively.

A part of the amino acid sequence of these components was obtained from human and bovine α3 and α4(IV) NC domain peptides [17, 38, 39]. Based on the few available amino acid sequences, degenerative oligo primers were made and partial cDNAs mainly covering the NC domain of the bovine α3(IV), α4(IV) and human α3(IV) were isolated by PCR amplification followed by library screening [40–42]. The complete primary structure of the human α3(IV) and α4(IV) chains has recently been determined [18, 25].

A striking feature of these two chains is that they are rich in cysteines. There are 24 cysteine residues in the α3(IV) chain, 4 more than the α1(IV) chain, 3 more than the α2(IV) chain. Five out of these are present in the collagenous domain but not in interruptions, characteristic for α3(IV) and α4(IV) chains. The α4(IV) chain contains 31 cysteine residues highest of all the type IV collagen chains. The cysteines are known to be important in the inter- or intra-chain cross-linking and they are conserved in most of the cases. Thirteen of the 31 cystines in α4(IV) chains are not conserved through all the type IV chains, 7 in the Gly-Xaa-Yaa triplets and 6 in the interruptions. The significance of the large number of cysteines in α4(IV) chain as well as the relative large number of that in α3(IV) chain is not yet known. It is possible that the type IV collagen molecules containing the α4(IV) chain forms a tighter network than those not containing this chain by the cross-links between the triple helices of α4(IV) chains or between the α4(IV) chains and other BM components. This hypothesis is more likely to be true for those six cysteines located in the Gly-Xaa-Yaa triplets. Some of the cysteines that are only present in the α3(IV) and α4(IV) chains such as the ones in the interruptions XVII implies that these cysteines may take part in the inter-chain cross-link of these two chains. Evidence of the type IV triple helical molecules containing [α3(IV)]₂α4(IV) heterotrimers and the colocalization of these two chains support this hypothesis. Both α3 and α4 chains are rich in cysteines and rich in GBM. It is possible that the highly cross-linked BM network resulted in by the rich content of cysteines is determined by the special function of GBM.

The α5(IV) and α6(IV) Chains
Both the α5(IV) and α6(IV) chains were discovered by molecular cloning (see table 2 for the major features of these two chains). The α5(IV) chain was isolated by library screening with an oligonucleotide containing sequences common to the NC1 domains of the α1 and α2 chain of human, mouse and *Drosophila* in the attempt to look for the α3 and α4 chains [43]. The complete primary structure of the human and mouse chain has been determined [44, Killen, pers. commun.]. There is one potential glycosylation site in the 7S domain with the sequence Asn-Gly-Try which agrees with the consensus sequence Asn-Xaa-Thr/Ser. The Asn in the corresponding position of the native molecules of α1(IV) and α2(IV) is glycosylated [22]. Differing from the α1(IV) and α2(IV) chains, no RGD triplet sequence can be found in α5(IV). The calculated molecular mass of the mature human α5(IV) chain is 158 kD.

The α6(IV) chain was first isolated based on the prediction that type IV collagen exists in pairs and that a gene located upstream of *COL4A5* may be responsible for part of the X-linked AS cases and AS-associated leiomyomatosis. Genomic fragments on the 5′ end of *COL4A5* were screened and a particular fragment was identified to contain coding sequences by cross-species hybridization. cDNA clones coding for the new chain were therefore identified [3]. Interestingly, unlike the other type IV collagen chains identified so far, the α6(IV) chain is encoded by two transcripts which differ at their extreme 5′ ends [5, 45]. These two transcripts are likely to be produced by two promoters since no evidence of alternative splicing was seen. These two transcripts, however, are predicted to produce an identical mature α6(IV) chain since only the sequences that encode the signal peptides are variable. The level of expression of these two transcripts seems to vary between tissues and cell lines [46, Zhao et al., unpubl. data]. Whether the presence of two signal peptides for the α6(IV) chain is required for tissue specific or developmentally regulated expression needs further study. However, it has recently been shown that, unlike the other paired collagen IV chains that are often colocalized, the α5(IV) and α6(IV) chains are differentially expressed in some tissues discussed at the end of this chapter.

Alternative Splicing of Type IV Collagens
Alternative splicing occurs in a number of collagen genes such as the α1 chain of the type II, XIII, and XVIII collagen, the α2 and α3 chains of type

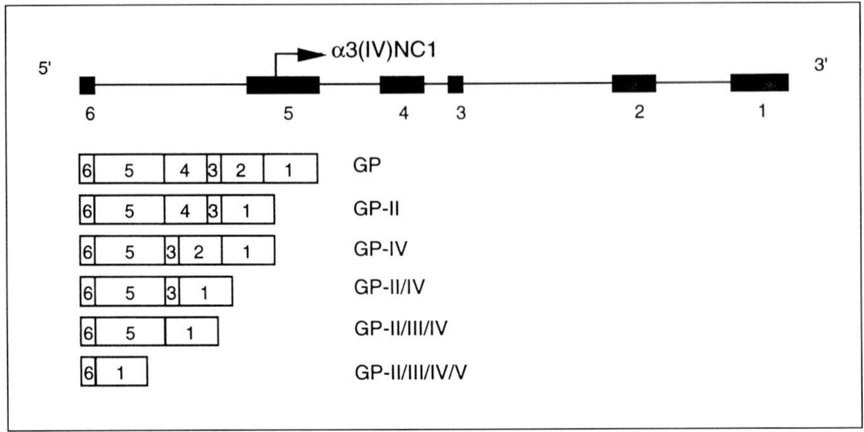

Fig. 4. Alternative splicing in the α3(IV) chain. *Top*: Schematic illustration of the 3′ end genomic structure of the COL4A3 gene. The exons (filled in boxes) are numbered from the 3′ end. *Bottom*: Various transcripts produced by alternative splicing events are indicated by open boxes labeled with corresponding exon numbers [47–49].

VI collagen. These alternative splicing events have been proposed to allow variation in the lengths of chains between different sites of expression. In the α1 and α2 chains of type I collagen, aberrant alternative splicing has been associated with osteogenesis imperfecta and Ehlers-Danlos syndrome type VII.

Analyses of the α1, α2, α4, and α6 chains of human type IV collagen have not revealed any alternative splicing events. Two groups have independently found alternatively spliced transcripts for the α3(IV) chain in human kidney samples from normals and Goodpasture syndrome patients [47–49]. In each case, the alernatively spliced transcripts lack one or more of the exon(s) encoding the NC1 domain: exon 2, exon 4, exons 2 and 4 together, exons 2, 3, and 4 together, and exons 2, 3, 4, 5 together (numbered from the 3′ end of the gene) (fig. 4). If translated, the exon 2-minus transcript (GP-II) would cause a frameshift resulting in an α3(IV) chain with a unique C-terminus. The exon 4-minus transcript (GP-IV) would produce a shortened α3(IV) chain whose NC1 domain lacks 183 residues including 11 of the 12 cysteine residues. The transcript lacking exons 2 through 5 (GP-II/III/IV/V) would have no NC1 domain at all. There is no evidence that these alternatively spliced α3(IV) transcripts are translated. If translation occurs, the novel peptides may play a role in the control of the amount of the mature functional α3(IV) chain. Indeed, Feng et al. [48] have found that the ratio between the full-length transcript and the spliced variants seems to vary with the stage of development.

The α5(IV) chain also seems to undergo alternative splicing. Saito et al. [50] recently reported the detection of an alternatively spliced α5(IV) chain in normal human adult kidney and skin samples. The alternatively spliced transcript lacks exon 50. This transcript, if translated, would result in a truncated α5(IV) chain lacking 84 C-terminal residues including 5 of the 12 cysteines in the NC1 domain. The relative abundance of this transcript – in relationship to the full-length transcript – was not determined. The significance of the presence of such transcripts remains to be determined.

The alternative splicing of type IV collagen is not unique to man. The alternative use of exons 9 or 10 and exons 108 or 111 have been described in the α2(IV) chain of nematodes *C. elegans* [10] and *A. suum* [51]. In both cases, one transcript is predominantly expressed in embryos while the other is predominantly expressed in adults; the alternative splicing may reflect the need of fast remodeling in embryos and the relatively stable supporting role of BMs in adults. One should note that the lengths of the α chains produced by alternative splicing in nematodes varies by only one amino acid.

Biosynthesis and Molecular Assembly of Type IV Collagen

Most studies of the protein chemistry of type IV collagens involve the abundant α1(IV) and α2(IV) chains. The biosynthesis of type IV collagen involves the same co- and posttranslational modifications as collageneous proteins in general [52]. These include the removal of the signal peptide in the endoplasmic reticulum; the hydroxylation of proline to 3-hydroxyproline in the Xaa-position; the hydroxylation of proline and lysine to 4-hydroxyproline and hydroxylysine, respectively, in the Yaa-position; glycosylation of hydroxylysine; the addition of a mannose-rich oligosacharide to the carboxyl propeptide; chain association, and helix formation. In contrast to fibrillar collagen, the posttranslational modification of type IV collagen does not involve the extracellular removal of propeptides by N- and C-proteinases.

Chain association begins in the NC1 domains of individual α chains. Residues that are not conserved between α chain isoforms are presumed to provide the specificity. The nature of these NC1 interactions is not yet clear but it has been shown that NC1 disulfide bonds were not formed at the time of association [53]. Triple helix formation proceeds from the carboxyl terminus towards the amino terminus in a zipper-like fashion [54]. As demonstrated for type I and III collagen, effective nucleation in type IV collagen presumably requires the right conformation of the NC domain disulfide bridges [55, 56].

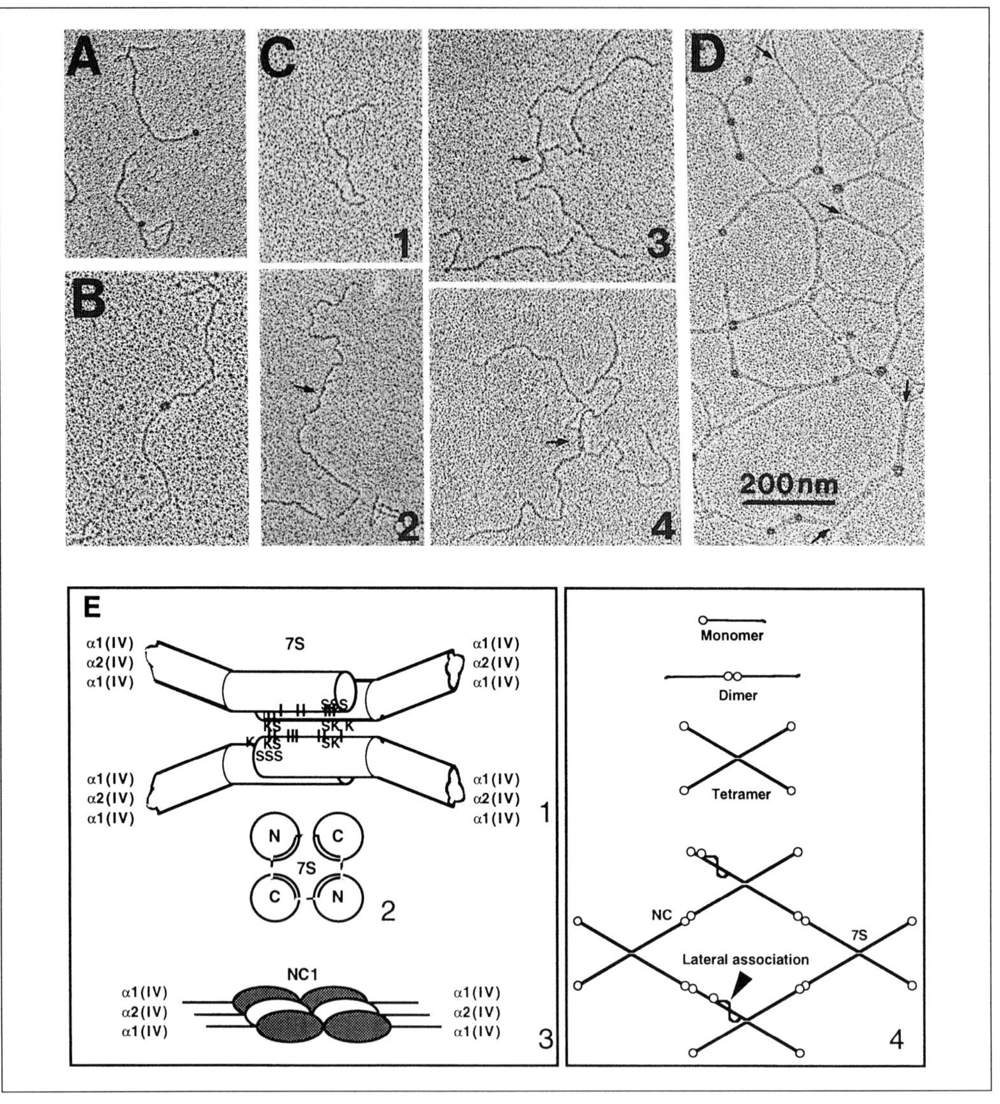

Fig. 5. Supermolecular assembly of type IV collagen network. *A–D* Electron micrograph of low-angle platinum/carbon rotary shadowed replicas of type IV collagen complexes [90]. *A* Monomers. *B* Two monomers associate into dimers via their NC domains. *C* Formation of 7S tetramers. The globular NC1 domains of these monomers have been proteolytically excised: (1) '7S' monomer; (2) '7S' antiparallel dimer; (3) '7S' trimer; (4) '7S' tetramer. *D* Polygonal network of type IV collagen is formed by lateral associations between dimers. The NC domains of dimers or free monomers can twist around the collagen helix and form so-called lateral associations. Schematic illustration of collagen IV network assembly is shown in *E4*. *E1–3* shows the N- and C-terminal interactions of type IV collagen molecules:

The assembly of triple helical molecules can be described as a three-step process: dimerization, tetramerization and lateral association (fig. 5). First, two triple-helical monomers form a linear dimer through the association of NC1 domains from two different molecules in the intracellular compartment. The assembly seems to be driven by inter-chain rearrangement of disulfide bonds between the cysteines of the NC1 domains resulting in a stable hexameric NC1 complex [57]. Homodimers of $\alpha1(IV)_2 \cdot \alpha2(IV)$ and $\alpha3(IV)_2 \cdot \alpha4(IV)$ [14, 34] and heterodimers $\alpha1(IV)_2 \cdot \alpha2(IV)$ with $\alpha3(IV)_2 \cdot \alpha4(IV)$ [14, 17, 58] have been described.

The second step involves four triple-helical monomers that form a spider-like tetramer through parallel and antiparallel interactions between their 7S domains [20, 59–62] (fig. 5). The tetramerization is first directed by hydrophobic interactions between the hydrophobic self-interacting edges on each 7S domain of the triple-helical molecule [22], and is then stabilized by lysine-derived nonreducible cross-links and disulfide bridges [22, 63]. The intermolecular disulfide bridges may be located in a symmetric or asymmetric manner and the lysine-derived non-reducible cross-links may be formed between noncollagenous sequence of one end of the chains and collagenous part of the antiparallel molecule, while they are located in the nonhelical region of both ends of the α chains in intersitial collagens [62] (fig. 5).

The hydrophobic self-interacting edge also determines the azimuthal orientation of the two antiparallel molecules, which is mainly contributed by $\alpha2(IV)$ not $\alpha1(IV)$. In the case of $\alpha1$ homotrimers, hypothetically only a symmetric rotationally interaction could take place instead of a distinct or axial one [22]. The location and structure as well as potential function of the asparagine-linked oligosaccharide units in the 7S domain have been explored [23, 24]. There are a total of 12 Asn-like oligosaccharides and 72 disaccharides in the 7S tetramer contributed by 12 α chains. All the N-linked oligosaccharides are located in the boundary of the overlap region and two thirds (48) of the disaccharides are located in the overlapping region, of which two thirds (32) are exposed on the surface of the triple-helix whereas the remaining 12 are buried inside. This asymmetrical distribution of carbohydrate units provide the protomer of each triple-helical molecule an amphipathic character and therefore contribute to stability of the tetramer except their steric hindrance

(1) N-terminal interactions between antiparallel monomers. SS = Disulfide cross-links; KK = putative lysine/hydroxylysine-derived cross-links [22, 63]. (2) Azimuthal orientation of the four antiparallel monomers at the 7S domain. Double lines indicate the interacting edges. C and N mark the orientation of the molecules. (3) C-terminal interaction of two monomers [modified from 57].

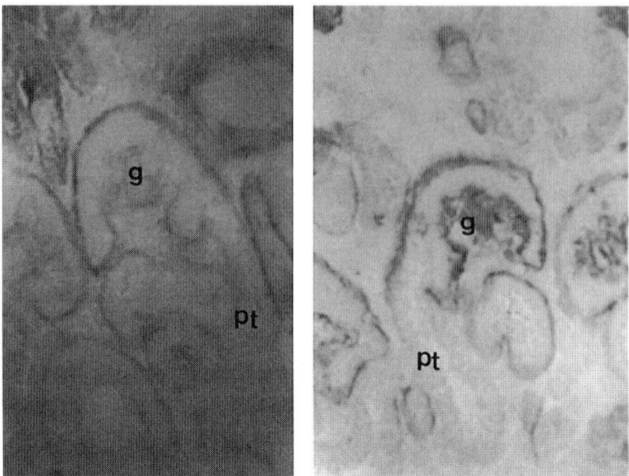

Fig. 6. Differential distribution of α5(IV) and α6(IV) chains in human 22-week fetal kidney. Sections are stained with (*A*) mouse anti-human α5(IV) antiserum and (*B*) rabbit anti-human α6(IV) antiserum using immunoperoxidase techniques. The differential expression of the α5(IV) and α6(IV) chains was observed at the early capillary loop stage (C- and U-shaped glomerulus). Neither α5(IV) nor α6(IV) was present in the proximal TBM. The α6(IV) chain was found in the Bowman's capsule and TBM. The α5(IV) chain was found in the Bowman's capsule, GBM and the TBM. pt = Proximal tubules, g = glomerulus [from 72, with permission].

role [23]. We do not yet know how the α3(IV) chain takes part in tetramerization since it does not have the highly conserved Asn-Xaa-Thr/ser glycosylation sequence.

In addition to the cross-linking at both ends of a type IV collagen monomer, the NC1 domain of a dimer or of free monomers can also bind to the triple-helical domain of neighboring molecules at intervals of about 100 nm. This process is so-called lateral association [64].

A loose, continuous network can be constructed in vitro through the above described interactions of both terminal regions of type IV collagen molecules (fig. 5). However, both in situ and in vitro studied networks have a polygonal irregular array appearance whose sides vary from one to several triple-helical thicknesses (fig. 6). This has been explained as the result of extensive lateral association of two or three molecules twisted together with an imperfect crossover periodicity of about 5 nm and the location of the linear dimerized carboxyl terminus at or near the polygonal vertices as often noticed [65]. Each dimer can have more than one spatial relationship with a neighboring dimer as well, which might be determined by the irregular spaced interruptions along the collagen chain [33]. It has been proposed, based on studies on

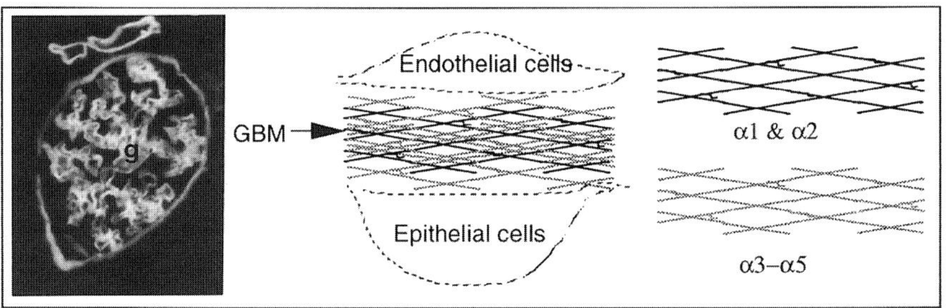

Fig. 7. Hypothetical GBM collagen IV networks. *Left*: GBM is recognized by a mouse anti-human α5(IV) chain-specific antibody. *Right*: Schematic illustration of the two independent type IV collagen networks in GBM. g = Glomerulus.

EHS tumor type IV collagen dimers, that the NC domain dimer formation occurs first, the lateral association second, and the amino-terminal tetramer formation last [65]. However, studies on tissue culture collagen monomers have indicated that the network formation is initiated by formation of the tetramer [60, 61]. The real order of reactions or first molecular recognition events in the self-assembly of type IV collagen is still unclear. The assembly of type IV collagen is illustrated in figure 5.

Tissue Distribution of the Type IV Collagen Chains

Type IV collagen chains exist only in BMs. The α1(IV) and α2(IV) chains are present in all BMs whereas the α3-α6(IV) chains have restricted tissue distribution (fig. 7).

Northern Analysis

Although we do not have complete knowledge of the tissue localization of all the type IV chains, all six chains are known to be expressed in the kidney, lung, skeletal muscle, meninges and heart by Northern studies [3, 5, 18, 25]. The expression patterns of all the chains are similar in adult and fetal heart, brain, lung, liver and kidney [5, 25]. The mRNA ratios of α3/α4 and α5/α6 were approximately constant in some tissues – fetal cerebellum, adult kidney, skeletal muscle and lung – but vary dramatically in other tissues. For example, α4(IV) mRNA is 5 times more abundant than α3(IV) in fetal kidney [18]; α6(IV) mRNA is 5 times more abundant than α5(IV) in adult heart [5]. The α3 and α4(IV) chains colocalize and the α5 and α6(IV) chains colocalize; in some tissues only α5 and α6(IV) are present – e.g. skin – whereas in other tissues – e.g. the thymus – only α3 and α4(IV) are present.

Striking differences in transcript size have been noted for the α4(IV) and α5(IV) chains. Two different α4(IV) transcripts ($\sim$7.5 and $\sim$10 kb) were detected in adult skeletal muscle and fetal and adult kidney and lung. A transcript of 6.5 kb was found in placenta [25]. Three different α5(IV) transcripts ($\sim$7.3, $\sim$7.0 and $\sim$5.0 kb) were detected in placenta [5]. These differences presumably represent the products of alternative splicing and/ or the use of alternative polyadenylation sites.

We need to be cautious in interpreting the Northern data since they (1) reflect only steady-state mRNA levels and (2) they do not reflect intraorgan variation in the expression of individual chains.

Immunochemical Studies of Tissue Distribution
Chain-specific antibodies have been produced against all six human isoforms. These antibodies have shown that the α3(IV) and α4(IV) chains are expressed in the human kidney, eye, cochlea, lung, brain, and synaptic muscle fibers [66–69]. The α5 and α6(IV) chains are expressed in the kidney, eye, lung, esophagus and skin [43, 70–72].

In the kidney, both α3(IV) and α4(IV) chains are expressed in the GBM, Bowman's capsule and distal tubules; the α5(IV) chain is also present at these sites and is also found in the collecting ducts [66, 72]. The α6(IV) colocalizes with the α5(IV) chain throughout the kidney but is notably absent from the GBM [72]. The differential expression of the α5(IV) and α6(IV) chains in the kidney is first detected at the capillary loop stage of kidney development when the glomeruli start to form (fig. 6) [72]. Differential localization of the α5(IV) and α6(IV) chains in the GBM indicates that the α5(IV) chain takes part in the GBM network as a homotrimer or as a heterotrimer with type IV collagen α chains other than the α6(IV) chain [72].

It has long been noted that sera from patients with Goodpasture syndrome fail to bind GBM from X-linked Alport patients. Immunohistochemical studies using chain-specific antibodies have shown that the α3(IV)-α5(IV) chains are absent in the GBM of many X-linked Alport patients [70, 73]. It has recently been shown that the α6(IV), along with the α3-α5(IV) chains, is absent from Bowman's capsules and tubules in these patients while the α1(IV) and α2(IV) chains are typically increased in the GBMs of AS kidneys [72]. The α6(IV) chain is present in patients with other renal diseases including thin BM disease. The mechanisms for the absence of α3, α4, and α6(IV) chains in the GBMs of these patients may be: (1) failure to incorporate these chains as part of the GBM or TBM into a stable heterotrimer due to the presence of a mutant α5(IV) chain; (2) abnormal transcription or translation of these chains due

Table 3. Tissue distributions of the six collagen IV chains in renal BMs from normal individuals and from patients with AS [43, 66–74, 96]

	α1	α2	α3	α4	α5	α6
Normal distribution						
GBM	$+/-$	$+/-$	$+$	$+$	$+$	$-$
Bowman's capsules	$+$	$+$	$+/-$ S	$+/-$	$+$	$+$
Proximal tubules	$+$	$+$	$-$	$-$	$-$	$-$
Henle's loop	$+$	$+$	$-$	$-$	$-$	$-$
Distal tubules	$+$	$+$	$+/-$	$+/-$	$+$	$+$
Collecting ducts						
Medullar	$+$	$+$	$-$	$-$	$+$	$+$
Cortical	$+$	$+$	$-$	$-$	$+$	$+$
Mesangium	$+$	$+$	$-$	$-$	$-$	$-$
Blood vessels	$+$	$+$	$-$	$-$	$-$	$-$
Skin	$+$	$+$	$-$	$-$	$+$	$+$
Anterior lens capsule	$+$	$+$	$-$	$-$	$-$	ND
X-linked AS (α5 mutations)						
GBM	$+$	$+$	$-$	$-$	$-$	$-$
Bowman's capsules	$+$	$+$	$-$	$-$	$-$	$-$
Proximal tubules	$+$	$+$	$-$	$-$	$-$	$-$
Henle's loop	$+$	$+$	$-$	$-$	$-$	$-$
Distal tubules	$+$	$+$	$-$	$-$	$-$	$-$
Collecting ducts						
Medullar	$+$	$+$	$-$	$-$	$-$	$-$
Cortical	$+$	$+$	$-$	$-$	$-$	$-$
Mesangium	$+$	$+$	$-$	$-$	$-$	$-$
Blood vessels	$+$	$+$	$-$	$-$	$-$	$-$
Skin	$+$	$+$	$-$	$-$	$-$	$-$
Anterior lens capsule	$+$	$+$	$-$	$-$	$-$	ND
Autosomal recessive AS (α3 or α4 mutations)						
GBM	$+$	$+$	$-$	$-$	$-$	$-$
Bowman's capsules	$+$	$+$	$-$	$-$	$+$	ND
Proximal tubules	$+$	$+$	$-$	$-$	$-$	$-$
Distal tubules	$+$	$+$	$-$	$-$	$-$	$-$
Collecting ducts	$+$	$+$	$-$	$-$	$+$	ND

ND = Not determined; $+$ = stained; $-$ = unstained; $+/-$ = weekly stained; S = segmental stained.

to mutations in the $\alpha5(IV)$ chain. The tissue distribution of type IV collagen α chains in renal BMs is summarized in table 3.

As discussed above, the six type IV isoforms can be divided into two structurally-related groups: the *α1-like* group, which includes the $\alpha1$, $\alpha3$ and $\alpha5(IV)$ chains, and the *α2-like* group, which includes the $\alpha2$, $\alpha4$, and $\alpha6(IV)$ chains. Recent studies of the expression patterns of these chains in normal development and in tissues from Alport patients allow us to speculate that there may be two independently regulated networks of type IV collagens: one for the abundant chains, the $\alpha1$ and $\alpha2(IV)$ chains and the other for the $\alpha3$-$\alpha6(IV)$ chains. This hypothesis is supported by the following facts: (1) the $\alpha1(IV)$ and $\alpha2(IV)$ chains are widely distributed in most if not all tissues; (2) the $\alpha1(IV)$ and $\alpha2(IV)$ chains appear earlier than the $\alpha3(IV)$–$\alpha6(IV)$ chains during kidney development [72, 74]; (3) in some AS patients with $\alpha5(IV)$ mutations, the $\alpha1$ and $\alpha2(IV)$ chains are overexpressed in the GBM while the $\alpha3(IV)$–$\alpha6(IV)$ chains are absent [70, 72, 73]. The $\alpha3(IV)$–$\alpha6(IV)$ chains may form independent networks at some sites. For example, recessive $\alpha3(IV)$ mutations lead to absence of the $\alpha4(IV)$ and $\alpha5(IV)$ chains from GBM. However, in Bowman's capsule BM, $\alpha5(IV)$ is present in these patients but $\alpha4(IV)$ is not.

Biological Roles of Type IV Collagen

Our knowledge of the discriminant functions of the type IV isoforms is limited to circumstantial evidence based on (1) tissue distribution, described above, and (2) the involvement of various 'minor' chains in disease processes. The latter includes the mutations in the $\alpha3(IV)$, $\alpha4(IV)$ and $\alpha5(IV)$ chains in X-linked and autosomal recessive AS, and deletion of the $\alpha5(IV)$ and $\alpha6(IV)$ genes in AS-associated leiomyomatosis, which are discussed in detail elsewhere in this book.

Mutations in the abundant and widely expressed $\alpha1(IV)$ and $\alpha2(IV)$ chains are likely to be embryonic lethal. Indeed, point mutations in *emb-9 of C. elegans* which change a glycine to glutamic acid in the $\alpha1(IV)$ chain cause temperature-sensitive lethality during embryogenesis [9]. Mutations in the $\alpha2(IV)$ chain have similar effects [75].

A number of in vitro studies on the $\alpha1(IV)$ and $\alpha2(IV)$ chains have shown that, like laminin, type IV collagen has the ability to bind integrins and adhere to epithelial cells, human embryo fibroblasts, human HT1080 fibrosarcoma cells [76], endothelial cells [77], neuron-like rat pheochromocytoma cells [78] and murine melanoma cells [79] through its collagenous [19, 79, 80] and/or NC domains [81]. Synthetic peptides, such as Hep-I, derived from the NC1 domain of α1(IV) (TAGSLRKFSTM), have been shown to have a number of functions such as self-association, heparin binding, cell binding and adhesion [82]. The role of type IV collagen in the peripheral nervous system has recently been further elucidated [83]. The NC1 domain of type IV collagen can promote neurite outgrowth via the mediation of α1β1 integrin [83]. The 7S and triple-helical domains have no effect. However, a major cell-binding site with affinity to integrins α1β1 and α2β1 has been located in the triple helix domains of α1(IV) and α2(IV), 100 nm away from the amino terminus and close to interruption IX [19]. The recognition sequence is composed of two aspartate residues within each α1(IV) chain and one arginine residue within the α2(IV) chain [84]. Miles et al. [85], however, recently showed that the triple-helical conformation of the type IV chains is not critical for cellular recognition.

Increasing knowledge of the sequences, genetics, and localizations of the six type IV collagen chains now allow us to further dissect the functions of these special collagenous proteins in normal development and in disease status including diabetic nephropathy, tumorigenesis and tumor metastasis, to study the genetic control of the expression of these genes, and to plan gene therapy.

References

1   Solomon E, Hall V, Kurkinen M: The human α2(IV) collagen gene, COL4A2, is syntenic with the α1(IV) gene, COL4A1, on chromosome 13. Ann Hum Genet 1987;51:125–127.
2   Mariyama M, Zheng K, Yang FT, Reeders ST: Colocalization of the genes for the α3(IV) and α4(IV) chains of type IV collagen to chromosome 2 bands q35–q37. Genomics 1992;13:809–813.
3   Zhou J, Mochizuki T, Smeets H, Antignac C, Laurila P, de PA, Tryggvason K, Reeders ST: Deletion of the paired α5(IV) and α6(IV) collagen genes in inherited smooth muscle tumors. Science 1993;261:1167–1169.
4   Hudson BG, Reeders ST, Tryggvason K: Type IV collagen: Structure, gene organization, and role in human diseases. Molecular basis of Goodpasture and Alport syndromes and diffuse leiomyomatosis (review). J Biol Chem 1993;268:26033–26036.
5   Zhou J, Ding M, Zhao Z, Reeders ST: Complete primary structure of the sixth chain of human basement membrane collagen, α6(IV). Isolation of the cDNAs for α6(IV) and comparison with five other type IV collagen chains. J Biol Chem 1994;269:13193–13199.
6   Boyd CD, Toth FS, Gadi IK, Litt M, Condon MR, Kolbe M, Hagen IK, Kurkinen M, Mackenzie JW, Magenis E: The gene coding for human pro α1(IV) collagen and pro α2(IV) collagen are both located at the end of the long arm of chromosome 13. Am J Hum Genet 1988;42:309–314.
7   Blumberg B, MacKrell AJ, Olson PF, Kurkinen M, Monson JM, Natzle JE, Fessler JH: Basement membrane procollagen IV and its specialized carboxyl domain are conserved in Drosophila, mouse, and human. J Biol Chem 1987;262:5947–5950.
8   Guo XD, Kramer JM: The two *Caenorhabditis elegans* basement membrane (type IV) collagen genes are located on separate chromosomes. J Biol Chem 1989;264:17574–17582.
9   Guo XD, Johnson JJ, Kramer JM: Embryonic lethality caused by mutations in basement membrane collagen of *Caenorhabditis elegans.* Nature 1991;349:707–709.
10 Sibley MH, Johnson JJ, Mello CC, Kramer JM: Genetic identification, sequence, and alternative splicing of the *Caenorhabditis elegans* α2(IV) collagen gene. J Cell Biol 1993;123:255–264.

11 Pettitt J, Kingston IB: The complete primary structure of a nematode $\alpha2$(IV) collagen and the partial structural organization of its gene. J Biol Chem 1991;266:16149–16156.

12 Exposito JY, D'Alessio M, Di LM, Ramirez F: Complete primary structure of a sea urchin type IV collagen α chain and analysis of the 5′ end of its gene. J Biol Chem 1993;268:5249–5254.

13 Exposito JY, Suzuki H, Geourjon C, Garrone R, Solursh M, Ramirez F: Identification of a cell lineage-specific gene coding for a sea urchin $\alpha2$(IV)-like collagen chain. J Biol Chem 1994;269:13167–13171.

14 Johansson C, Burkowski R, Wieslander J: The structural organization of type IV collagen. Identification of three NC1 populations in the glomerular basement membrane. J Biol Chem 1992;267:24533–24537.

15 Kleppel MM, Fan W, Cheong HI, Michael AF: Evidence for separate networks of classical and novel basement membrane collagen. Characterization of $\alpha3$(IV)-Alport antigen heterodimer. J Biol Chem 1992;267:4137–4142.

16 Derry CJ, Pickering M, Baker C, Pusey CD. Identification of the Goodpasture antigen, $\alpha3$(IV) NC1, and four other NC1 domains of type IV collagen, by amino-terminal sequence analysis of human glomerular basement membrane separated by two-dimensional electrophoresis. Exp Nephrol 1994;2:249–256.

17 Saus J, Wieslander J, Langeveld JP, Quinones S, Hudson BG: Identification of the Goodpasture antigen as the $\alpha3$(IV) chain of collagen IV. J Biol Chem 1988;263:13374–13380.

18 Leinonen A, Mariyama M, Mochizuki T, Tryggvason K, Reeders ST: Complete primary structure of the human type IV collagen $\alpha4$(IV) chain. Comparison with structure and expression of the other α(IV) chains. J Biol Chem 1994;269:26172–26177.

19 Vandenberg P, Kern A, Ries A, Luckenbill EL, MannK, Kuhn K: Characterization of a type IV collagen major cell binding site with affinity to the $\alpha1\beta1$ and the $\alpha2\beta1$ integrins. J Cell Biol 1991;113:1475–1483.

20 Timpl R, Wiedemann H, van DV, Furthmayr H, Kuhn K: A nework model for the organization of type IV collagen molecules in basement membranes. Eur J Biochem 1981;120:203–211.

21 Risteli J, Schuppan D, Glanville RW, Timpl R: Immunochemical distinction between two different chains of type IV collagen. Biochem J 1980;191:517–522.

22 Siebold B, Qian RA, Glanville RW, Hofmann H, Deutzmann R, Kuhn K: Construction of a model for the aggregation and cross-linking region (7S domain) of type IV collagen based upon an evaluation of the primary structure of the $\alpha1$ and $\alpha2$ chains in this region. Eur J Biochem 1987;168:569–575.

23 Nayak BR, Spiro RG: Localization and structure of the asparagine-linked oligosaccharides of type IV collagen from glomerular basement membrane and lens capsule. J Biol Chem 1991;266:13978–13987.

24 Langeveld JP, Noelken ME, Hard K, Todd P, Vliegenthart JF, Rouse J, Hudson BG: Bovine glomerular basement membrane. Location and structure of the asparagine-linked oligosaccharide units and their potential role in the assembly of the 7S collagen IV tetramer. J Biol Chem 1991;266:2622–2631.

25 Mariyama M, Leinonen A, Mochizuki T, Tryggvason K, Reeders ST: Complete primary structure of the human $\alpha3$(IV) collagen chain. Coexpression of the $\alpha3$(IV) and $\alpha4$(IV) collagen chains in human tissues. J Biol Chem 1994;269:23013–23017.

26 Timple R, Aumailley M: Biochemistry of basement membranes. Adv Nephrol 1989;18:59–76.

27 Oberbaumer I, Laurent M, Schwarz U, Sakurai Y, Yamada Y, Vogeli G, Voss T, Siebold B, Glanville RW, Kuhn K: Amino acid sequence of the non-collagenous globular domain (NC1) of the $\alpha1$(IV) chain of basement membrane collagen as derived from complementary DNA. Eur J Biochem 1985;147:217–224.

28 Muthukumaran H, Blumberg B, Kurkinen M: The complete primary structure for the $\alpha1$-chain of mouse collagen IV. Differential evolution of collagen IV domains. J Biol Chem 1989;264:6310–6317.

29 Brazel D, Pollner R, Oberbaumer I, Kuhn K: Human basement membrane collagen (type IV). The amino acid sequence of the $\alpha2$(IV) chain and its comparison with the $\alpha1$(IV) chain reveals deletions in the $\alpha1$(IV) chain. Eur J Biochem 1988;172:35–42.

30   Kaytes PS, Theriault NY, Vogeli G: Homologies between the non-collagenous C-terminal (NC1) globular domains of the α1 and α2 subunits of type-IV collagen. Gene 1987;54:141–146.

31   Hostikka SL, Tryggvason K: The complete primary structure of the α2 chain of human type IV collagen and comparison with the α1(IV) chain. J Biol Chem 1988;263:19488–19493.

32   Babel W, Glanville RW: Structure of human-basement-membrane (type IV) collagen. Complete amino-acid sequence of a 914-residue-long pepsin fragment from the α1(IV) chain. Eur J Biochem 1984;143:545–556.

33   Yurchenco PD, Ruben GC: Basement membrane structure in situ: Evidence for lateral associations in the type IV collagen network. J Cell Biol 1987;2559–2568.

34   Butkowski RJ, Wieslander J, Wisdom BJ, Barr JF, Noelken ME, Hudson BG: Properties of the globular domain of type IV collagen and its relationship to the Goodpasture antigen. J Biol Chem 1985;260:3739–3747.

35   Wieslander J, Langeveld J, Butkowski R, Jodlowski, M, Noelken M, Hudson BG: Physical and immunochemical studies of the globular domain of type IV collagen. Cryptic properties of the Goodpasture antigen. J Biol Chem 1985;260:8564–8570.

36   Kleppel MM, Michael AF, Fish AJ: Antibody specificity of human glomerular basement membrane type IV collagen NC1 subunits. Species variation in subunit composition. J Biol Chem 1986;261: 16547–16552.

37   Butkowski RJ, Langeveld JP, Wieslander J, Hamilton J, Hudson BG: Localization of the Goodpasture epitope to a novel chain of basement membrane collagen. J Biol Chem 1987;262:7874–7877.

38   Butkowski RJ, Shen GQ, Wieslander J, Michael AF, Fish AJ: Characterization of type IV collagen NC1 monomers and Goodpasture antigen in human renal basement membranes. J Lab Clin Med 1990;115:365–373.

39   Gunwar S, Saus J, Noelken ME, Hudson BG: Glomerular basement membrane. Identification of a fourth chain, α4, of type IV collagen. J Biol Chem 1990;265:5466–5469.

40   Morrison KE, Germino GG, Reeders ST: Use of the polymerase chain reaction to clone and sequence a cDNA encoding the bovine α3 chain of tyope IV collagen. J Biol Chem 1991;266:34–39.

41   Morrison KE, Mariyama M, Yang FT, Reeders ST: Sequence and localization of a partial cDNA encoding the human α3 chain of type IV collagen. Am J Hum Genet 1991;49:545–554.

42   Mariyama M, Kalluri R, Hudson BG, Reeders ST: The α4(IV) chain of basement membrane collagen. Isolation of cDNAs encoding bovine α4(IV) and comparison with other type IV collagens. J Biol Chem 1992;267:1253–1258.

43   Hostikka SL, Eddy RL, Byers MG, Hoyhtya M, Shows TB, Tryggvason K: Identification of a distinct type IV collagen α chain with restricted kidney distribution and assignment of its gene to the locus of X chromosome-linked Alport syndrome. Proc Natl Acad Sci USA 1990;87:1606–1610.

44   Zhou J, Hertz JM, Leinonen A, Tryggvason K: Complete amino acid sequence of the human α5(IV) collagen chain and identification of a single-base mutation in exon 23 converting glycine 521 in the collagenous domain to cysteine in an Alport syndrome patient. J Biol. Chem 1992;267:12475–12481.

45   Oohashi T, Sugimoto M, Mattei MG, Ninomiya Y: Identification of a new collagen IV chain, α6(IV), by cDNA isolation and assignment of the gene to chromosome Xq22, which is the same locus for COL4A5. J Biol Chem 1994;269:7520–7526.

46   Sugimoto M, Oohashi T, Ninomiya Y: The genes COL4A5 and COL4A6, coding for basement membrane collagen chains α5(IV) and α6(IV), are located head-to-head in close proximity on human chromosome Xq22 and COL4A6 is transcribed from two alternative promoters. Proc Natl Acad Sci USA 1994;91:11679–11683.

47   Bernal D, Quinones S, Saus J: The human mRNA encoding the Goodpasture antigen is alternatively spliced. J Biol Chem 1993;268:12090–12094.

48   Feng L, Xia Y, Wilson CB: Alternative splicing of the NC1 domain of the human α3(IV) collagen gene. Differential expression of mRNA transcripts that predict three protein variants with distinct carboxyl regions. J Biol Chem 1994;269:2342–2348.

49   Penades JR, Bernal D, Revert F, Johansson C, Fresquet VJ, Cervera J, Wieslander J, Quinones S, Saus J: Characterization and expression of multiple alternatively spliced transcripts of the Goodpas-

ture antigen gene region. Goodpasture antibodies recognize recombinant proteins representing the autoantigen and one of its alternative forms. Eur J Biochem 1995;229:754–760.

50   Saito A, Sakatsume M, Yamazaki H, Arakawa M: Alternative splicing in the α5(IV) collagen gene in human kidney and skin tissues. Nippon Jinzo Gakkai Shi 1994;36:19–24.

51   Pettitt J, Kingston IB: Developmentally regulated alternative splicing of a nematode type IV collagen gene. Dev Biol 1994;161:22–29.

52   Kivirikko KI, Myllyla R: Recent developments in posttranslational modification: Intracellular processing. Methods Enzymol 1987;144:96–114.

53   Fessler LI, Fessler JH: Identification of the carboxyl peptides of mouse procollagen IV and its implications for the assembly and structure of basement membrane procollagen. J Biol Chem 1982; 257:9804–9810.

54   Dolz R, Engel J, Kuhn K: Folding of collagen IV. Eur J Biochem 1988;178:357–366.

55   Bachinger HP, Bruckner P, Timpl R, Prockop DJ, Engel J: Folding mechanism of the triple helix in type-III collagen and type-III pN-collagen. Role of disulfide bridges and peptide bond isomerization. Eur J Biochem 1980;106:619–632.

56   Bruckner P, Eikenberry EF: Formation of the triple helix of type I procollagen in cellulo. Temperature-dependent kinetics support a model based on cis in equilibrium trans isomerization of peptide bonds. Eur J Biochem 1984;140:391–395.

57   Siebold B, Deutzmann R, Kuhn K: The arrangement of intra- and intermolecular disulfide bonds in the carboxyterminal, noncollagenous aggregation and cross-linking domain of basement-membrane type IV collagen. Eur J Biochem 1988;176:617–624.

58   Langeveld JP, Wieslander J, Timoneda J, McKinney P, Butkowski RJ, Wisdom BJ, Hudson BG: Structural heterogeneity of the noncollagenous domain of basement membrane collagen. J Biol Chem 1988;263:10481–10488.

59   Backinger HP, Doege KJ, Petschek JP, Fessler LI, Fessler JH: Structural implications from an electron microscopic comparison of procollagen V with procollagen I, pC-collagen I, procollagen IV, and a Drosophila procollagen. J Biol Chem 1982;257:14590–14592.

60   Duncan KG, Fessler LI, Bachinger HP, Fessler JH: Procollagen IV. Association to tetramers. J Biol Chem 1983;258:5869–5877.

61   Oberbaumer I, Wiedemann H, Timpl R, Kuhn K: Shape and assembly of type IV procollagen obtained from cell culture. EMBO J 1982;1:805–810.

62   Kuhn K, Wiedemann H, Timple R, Risteli J, Dieringer H, Voss T, Glanville RW: Macromolecular structure of basement membrane collagens. FEBS Lett 1981;125:123–128.

63   Glanville RW, Qian RQ, Siebold B, Risteli J, Kuhn K: Amino acid sequence of the N-terminal aggregation and cross-linking region (7S domain) of the α1(IV) chain of human basement membrane collagen. Eur J Biochem 1985;152:213–219.

64   Tsilibary EC, Charonis AS: The role of the main noncollagenous domain (NC1) in type IV collagen self-assembly. J Cell Biol 1986;103:2467–2473.

65   Yurchenco PD, Furthmayr H: Self-assembly of basement membrane collagen. Biochemistry 1984; 23:1839–1850.

66   Kleppel MM, Kashtan C, Santi PA, Wieslander J, Michael AF: Distribution of familial nephritis antigen in normal tissue and renal basement membranes of patients with homozygous and heterozygous Alport familial nephritis. Relationship of familial nephritis and Goodpasture antigens to novel collagen chains and type IV collagen. Lab Invest 1989;61:278–289.

67   Kleppel MM, Santi PA, Cameron JD, Wieslander J, Michael AF: Human tissue distribution of novel basement membrane collagen. Am J Pathol 1989;134:813–825.

68   Butkowski RJ, Wieslander J, Kleppel M, Michael AF, Fish AJ: Basement membrane collagen in the kidney: Regional localization of novel chains related to collagen IV. Kidney Int 1989;35: 1195–1202.

69   Sanes JR, Engvall E, Butkowski R, Hunter DD: Molecular heterogeneity of basal laminae: Isoforms of laminin and collagen IV at the neuromuscular junction and elsewhere. J Cell Biol 1990;111: 1685–1699.

70   Kleppel MM, Fan WW, Cheong HI, Kashtan CE, Michael AF: Immunochemical studies of the Alport antigen. Kidney Int 1992;41:1629–1637.

71   Yoshioka K, Hino S, Takemura T, Maki S, Wieslander J, Takekoshi Y, Makino H, Kagawa M, Sado Y, Kashtan CE: Type IV collagen α5 chain. Normal distribution and abnormalities in X-linked Alport syndrome revealed by monoclonal antibody. Am J Pathol 1994;144:989–996.

72   Peissel B, Geng L, Kalluri R, Kashtan K, Rennke HG, Gallo GR, Sun MJ, Hudson BG, Neilson EG, Zhou J: Comparative distribution of the α1(IV), α5(IV) and α6(IV) collagen chains in normal human adult and fetal tissues and in kidneys from X-linked Alport syndrome patients. J Clin Invest 1995;96:1948–1957.

73   Kashtan CE, Kim Y: Distribution of the α1 and α2 chains of collagen IV and of collagens V and VI in Alport syndrome. Kidney Int 1992;42:115–126.

74   Miner JH, Sanes JR: Collagen IV α3, α4, and α5 chains in rodent basal laminae: Sequence, distribution, association with laminins, and developmental switches. J Cell Biol 1994;127:879–891.

75   Sibley MH, Graham PL, von Mende N, Kramer JM: Mutations in the α2(IV) basement membrane collagen gene of *Caenorhabditis elegans* produce phenotypes of differing severities. EMBO J 1994;13:3278–3285.

76   Aumailley M, Timpl R: Attachment of cells to basement membrane collagen type IV. J Cell Biol 1986;103:1569–1575.

77   Herbst TJ, McCarthy JB, Tsilibary EC, Furcht LT: Differential effects of laminin, intact type IV collagen, and specific domains of type IV collagen on endothelial cell adhesion and migration. J Cell Biol 1988;106:1365–1373.

78   Tomaselli KJ, Damsky CH, Reichardt LF: Interactions of a neuronal cell line (PC12) with laminin, collagen IV, and fibronectin: Identification of integrin-related glycoproteins involved in attachment and process outgrowth. J Cell Biol 1987;105:2347–2358.

79   Syfrig J, Mann K, Paulsson M: An abundant chick gizzard integrin is the avian α1β1 integrin heterodimer and functions as a divalent cation-dependent collagen IV receptor. Exp Cell Res 1991;194:165–173.

80   Chelberg MK, McCarthy JB, Skubitz AP, Furcht LT, Tsilibary EC: Characterization of a synthetic peptide from type IV collagen that promotes melanoma cell adhesion, spreading, and motility. J Cell Biol 1990;111:261–270.

81   Chelberg MK, Tsilibary EC, Hauser AR, McCarthy JB: Type IV collagen-mediated melanoma cell adhesion and migration: Involvement of multiple, distinct domains of the collagen molecule. Cancer Res 1989;49:4796–4802.

82   Tsilibary EC, Reger LA, Vogel AM, Koliakos GG, Anderson SS, Charonis AS, Alegre JN, Furcht LT: Identification of a multifunctional, cell-binding peptide sequence from the α1(NC1) of type IV collagen. J Cell Biol 1990;111:1583–1591.

83   Lein PJ, Higgins D, Turner DC, Flier LA, Terranova VP: The NC1 domain of type IV collagen promotes axonal growth in sympathetic neurons through interaction with the α1β1 integrin. J Cell Biol 1991;113:417–428.

84   Eble JA, Golbik R, Mann K, Kuhn K: The α1β1 integrin recognition site of the basement membrane collagen molecule [α1(IV)]2 α2(IV). EMBO J 1993;12:4795–4802.

85   Miles AJ, Skubitz AP, Furcht LT, Fields GB: Promotion of cell adhesion by single-stranded and triple-helical peptide models of basement membrane collagen α1(IV)531–543. Evidence for conformationally dependent and conformationally independent type IV collagen cell adhesion sites. J Biol Chem 1994;269:30939–30945.

86   Soininen R, Haka RT, Prockop DJ, Tryggvason K: Complete primary structure of the α1-chain of human basement membrane (type IV) collagen. FEBS Lett 1987;225:188–194.

87   Brazel D, Oberbaumer I, Dieringer H, Babel W, Glanville RW, Deutzmann R, Kuhn K: Completion of the amino acid sequence of the α1 chain of human basement membrane collagen (type IV) reveals 21 non-triplet interruptions located within the collagenous domain. Eur J Biochem 1987;168:529–536.

88   Saus J, Quinones S, MacKrell A, Blumberg B, Muthukumaran G, Pihlajaniemi T, Kurkinen M: The complete primary structure of mouse α2(IV) collagen. Alignment with mouse α1(IV) collagen. J Biol Chem 1989;264:6318–6324.

89   Blumberg B, MacKrell AJ, Fessler JH: Drosophila basement membrane procollagen α1(IV). II. Complete cDNA sequence, genomic structure, and general implications for supramolecular assemblies. J Biol Chem 1988;263:18328–18337.
90   Yurchenco PD, Schittny JC: Molecular architecture of basement membranes. FASEB J 1990;4: 1577–1590.
91   Schwarz U, Schuppan D, Oberbaumer I, Glanville RW, Deutzmann R, Timple R, Kuhn K: Structure of mouse type IV collagen. Amino-acid sequence of the C-terminal 511-residue-long triple-helical segment of the α2(IV) chain and its comparison with the α1(IV) chain. Eur J Biochem 1986;157: 49–56.
92   Vogeli G, Horn E, Carter J, Kaytes PS: Proposed alignment of helical interruptions in the two subunits of the basement membrane (type IV) collagen. FEBS Lett 1986;206:29–32.
93   Blumberg B, MacKrell AJ, Olson PF, Kurkinen M, Monson JM, Natzle JE, Fessler JH: Basement membrane procollagen IV and its specialized carboxyl domain are conserved in Drosophila, mouse, and human. J Biol Chem 1987;262:5947–5950.
94   Cecchini JP, Knibiehler B, Mirre C, Le PY: Evidence for a type-IV-related collagen in *Drosophila melanogaster.* Evolutionary constancy of the carboxyl-terminal noncollagenous domain. Eur J Biochem 1987;165:587–593.
95   Killen PD, Francomano CA, Yamada Y, Modi WS, O'Brien SJ: Partial structure of the human α2(IV) collagen chain and chromosomal localization of the gene (COL4A2). Hum Genet 1987;77: 318–324.
96   Gubler MC, Knebelmann B, Beziau A, Broyer M, Pirson Y, Haddoum F, Kleppel MM, Antignac C: Autosomal recessive Alport syndrome: Immunohistochemical study of the type IV collagen chain distribution. Kidney Int 1995;47:1142–1147.

Jing Zhou, MD, PhD, Renal Division, Department of Medicine, Brigham and Women's Hospital, Harvard Medical School, Boston, MA 02115 (USA)

Tryggvason K (ed): Molecular Pathology and Genetics of Alport Syndrome.
Contrib Nephrol. Basel, Karger, 1996, vol 117, pp 105–129

The Type IV Collagen Gene Family

Pirkko Heikkilä, Raija Soininen

Biocenter Oulu and Department of Biochemistry, University of Oulu, Finland

Collagens are a family of structural proteins that form the major extracellular elements of the body. They are composed of three identical or similar α chains characterized by the repeating -Gly-X-Y- sequence where X and Y can be any amino acid. The glycine residue as every third amino acid makes possible the assembly of three chains into a triple-helical structure. Type IV collagen is one of major structural components of basement membranes where it forms a network providing scaffolding for other components and has an important role in the interaction of basement membranes with cells.

In type IV collagen, the α chains consist of three domains: an aggregation and cross-link area (called 7S) at the aminoterminus with several cysteine and lysine residues, a long collagenous region (but containing up to 26 interruptions in the repeating -Gly-X-Y- sequence), and at the carboxyterminus a globular noncollagenous (NC1) domain. The NC1 domain is the most conserved region between the different α(IV) chains as well as between different species [1, 2]. It contains six cysteine residues in invariant positions capable of forming three disulfide bonds leading to a symmetrical folding pattern. (For a detailed description of the structure of collagen IV chains, see chapter 4.)

Six different α(IV) chains have been identified. The most abundant chains are the $\alpha1$(IV) and the $\alpha2$(IV) chains that are found in almost all basement membranes forming a molecule composed of two $\alpha1$(IV) chains and one $\alpha2$(IV) chain. The other α chains, $\alpha3$(IV), $\alpha4$(IV), $\alpha5$(IV) and $\alpha6$(IV) have a much more restricted expression in respect of both time and location. Characterization of the primary structure of different α(IV) chains has revealed that they are very similar in structure, and that they can also be grouped into two classes, '$\alpha1$-like' and '$\alpha2$-like' (fig. 1), which points to a common ancestor gene that was

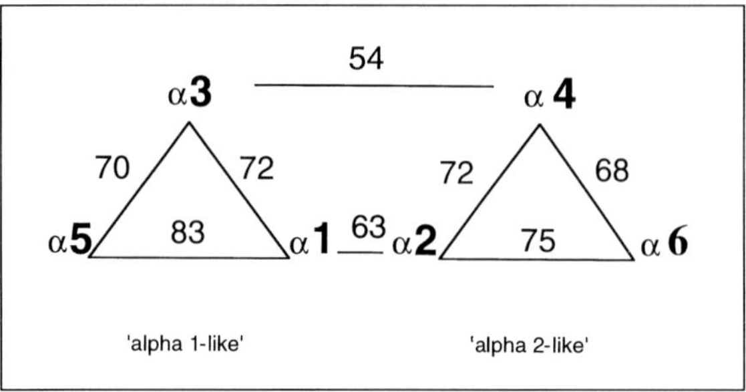

Fig. 1. Similarities in amino acid sequence of NC1 domains in different α(IV) chains. Percentage of identical or similar amino acids between different chains [3] is shown above the connecting line.

duplicated, followed by duplication of the duplicates. In the most conserved region, the NC1 domain, human α1(IV) and α5(IV) chain have 83% similarity in the amino acid sequence, and even the most divergent chains, α3(IV) and α4(IV), have 54% similarity in the primary structure of the NC1 domain [3].

Knowledge about the structure and regulation of human genes has proven indispensable for understanding the evolution, mechanisms of tissue development and regeneration, as well as the development of diseases. The study of type IV collagen genes has revealed a gene structure different from the previously characterized collagen genes, a unique bidirectional promoter organization, and mutations causing human diseases.

Chromosomal Localization of Type IV Collagen Genes

The first mammalian type IV collagen gene to be analyzed in detail and localized was the human *COL4A1* gene coding for the α1(IV) chain. It was localized to human chromosome 13 [4, 5], the location being assigned to the 13q34 region at the terminus of the chromosome [6, 7]. Soon thereafter the gene for the α2(IV) chain was localized to the same region as the α1(IV) chain gene on chromosome 13 [8–10]. This was the first demonstration of a close linkage between two genes coding for subunits of the same collagen type. In earlier reports it had been suggested that collagen genes are dispersed in the genome to prevent intergenic recombination between collagenous sequences [5]. On the other hand, it was speculated that possible additional genes coding for type IV collagen (which were not yet isolated at that time) would also be clustered in the same 13q33-34 region [8], a situation similar to other large multigene families, e.g. β-globin or histone genes, but later observations did not confirm this hypothesis. Pulsed-field gel electrophoresis was used to construct a macrorestriction map for the region covering the α1(IV) and α2(IV) genes [11] and showed that the two genes are localized within a 340-kb region, and that the *COL4A1* gene is between 90 and 200 kb in size, the size of *COL4A2* being of the same order. In spite of the close proximity, a high recombination frequency between the two genes was observed. The explanation for the increase of recombination was suggested to be either the presence of a recombination hot spot in the region or location in the terminal region of the chromosome [12].

The gene encoding the human α3(IV) chain, *COL4A3*, was localized to chromosome 2q35-37 [13], and later the gene encoding the α4(IV) chain, *COL4A4*, was assigned to the same region [14, 15], a situation similar to the α1(IV)-α2(IV) genes. Earlier, the gene for the α3 chain of type VI collagen, *COL6A3*, was mapped to the same region on chromosome 2 [16]. Several other genes coding for extracellular matrix proteins have been assigned to the long arm of chromosome 2, e.g. *COL3A1* and *COL5A2* on 2q24.3-2q31 [17]. The degree of synteny between the long arm of human chromosome 2 and proximal portion of mouse chromosome 1 was investigated, and it was found that *COL3A1* and *COL6A3* are border markers that define the limits of the syntenic chromosome segment [18]. In addition, the order of genes on mouse chromosome 1 and their human homologs on chromosome 2q appears to be conserved. It is not known yet if the genes coding for the two type IV collagen chains located close to *COL6A3* belong to the same syntenic segment because the order of the genes in this region is not known.

The gene encoding the α5(IV) chain was assigned to X chromosome region Xq22, the locus previously assigned to the X-linked form of Alport syndrome [19, 20], and the murine Col4a5 to mouse chromosome X linkage group F, the most extensively conserved linkage group between mouse and human X chromosomes [21]. The gene for the α6(IV) chain was assigned to the same region by DNA cloning and sequencing [22].

In the nematode *Caenorhabditis elegans*, two genes encoding type IV collagen α chains, *clb-1*, the product of which is most similar to the mammalian α2(IV) chain, and *clb-2* encoding an α1(IV)-like chain, have been identified [2], proving that the primordial type IV collagen gene has duplicated prior to the separation of the vertebrate and invertebrate lines. The genes are located on separate chromosomes: *clb-1* on chromosome X and *clb-2* on chromosome III [2]. In contrast, only one type IV collagen gene has been found in *Drosophila*, the molecule being a homotrimer [1], so it is likely that one of the genes was lost during the evolution of the *Drosophila* species.

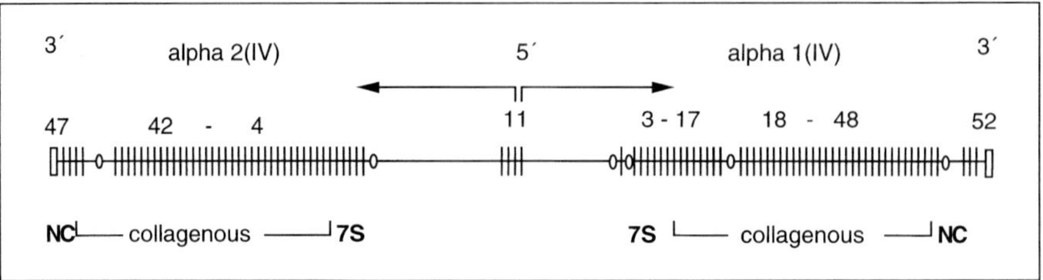

Fig. 2. COL4A1-COL4A2 locus. Exons are depicted as bars and numbered, introns as a horizontal line. Sizes are not in scale. Gaps in published maps are depicted as circles. Below, regions in the gene product encoded by respective exons.

Exon-Intron Structure of Type IV Collagen Genes

Characterization of the first genomic clones containing parts of type IV collagen genes from man and mouse [23–26] already revealed some typical features for this collagen gene class: The sizes of the exons do not appear to follow the 54-bp-unit size rule characteristic for fibrillar collagen genes. Furthermore, the sizes of introns are large, resulting in very large size of the type IV collagen genes. By macrorestriction mapping it has been estimated that the size of the human $\alpha1$(IV) gene is 100–160 kb [11], the human $\alpha5$(IV) gene is 230–250 kb, and the human $\alpha6$(IV) gene is at least 425 kb [27]. The exon sizes are fully conserved between human and mouse genes.

The Gene Encoding α1 Chain of Type IV Collagen
The complete exon-intron structure (fig. 2, table 1) was first characterized from the human $\alpha1$(IV) gene [28]. The gene contains 52 exons, and the sizes of the exons in the collagenous region vary from 27 to 192 bp (table 1). About one half of the -Gly-X-Y- repeat coding exons start with the second base for the codon of glycine, the rest (with two exceptions) with a complete glycine codon. The unsplit codon is more common at the 5' end of the gene: the first 19 exons start with a complete codon. Sequences at the boundaries between exons and introns follow the general consensus rule (the intron starts with GT and ends with AG), except at the 3' boundary of exon 34 where the intron starts with GCAACGTCC. Here there is GC instead of GT in an otherwise complete consensus sequence. The end of this intron contains the normal AG sequence.

In the 3' end, the last exon of the gene contains 79 bp coding for the most carboxyterminal end of the protein, and 1,383 bp untranslated sequence with four polyadenylation signals. There are five exons coding for the globular

Table 1. Exons in α1(IV), α2(IV) and α5(IV) chain genes [28, 35, 38]

Exon No.	α1(IV) exon size bp	split codon at 5' end	α5(IV) exon size bp	split codon at 5' end	α2(IV) exon size bp	split codon at 5' end
					(244)	
1	(129)+84		(202)+81		(44)+44	
2	60		60		55	
3	90		90		81	
					135	
4	45		45			
5	45		45		45	
6	63		63			
					117	
7	54		54			
8	27		27		72*	
9	84*		81*		36	
10	63		63		63	
11	36		36			
					78*	
12	42*		42*			
13	87*		93*		93*	
14	27		54		36	
15	51*		57*		51*	
16	45		45		45	
17	54		54		54	
18	42		42		67*	
19	85*					
			133*			
20	36	+			111*	+
21	165*	+	174*	+	150*	+
22	96*	+	84*	+	92*	+
23	84*	+	93*	+	155*	+
24	71	+	71	+		
25	192*		192*		73*	
26	169*		169*		107	+
					202*	+
					60*	+
					57*	+
27	93*	+	93*	+	108	+
28	105*	+	105*	+	222*	+
29	98	+	98	+	162*	+
30	151*	+	151*	+	171	+
31	114*	+	114*	+	144	+

Table 1 (continued)

Exon No.	α1(IV)		α5(IV)		α2(IV)	
	exon size bp	split codon at 5' end	exon size bp	split codon at 5' end	exon size bp	split codon at 5' end
32	168*	+	168*	+	123*	+
33	90	+	90*	+	182*	+
34	153*	+	150*	+	64	
35	99	+	99	+	75*	+
36	90	+	90	+	108	+
37	140*	+	140*	+	108	+
38	127		127		72	+
39	81	+	81	+	126*	+
40	99	+	99	+	117*	+
41	51*	+	51*	+	162*	+
42	186*	I	186	+	99	+
			(18)	+		
43	134	+	134	+	147	+
44	73		73		117	+
45	72	+	72	+		
46	129*	+	129	+		
47	99	+	99	+		
48	213	+	213	+	192	+
49	178	+	178	+		
					287	+
50	115	+	115	+		
51	173		173			
					255+	
52	79+(1304)		79+(1166)		(833)	

Numbering of exons in the α1(IV) gene is shown. Untranslated sequences are in parentheses. Exons with interruptions in the -Gly-X-Y- repeat sequence are marked with an asterisk.

NC1 domain, and the fifth exon from the 3' end (exon 48) which is 213 bp in size, is a junction exon that contains sequences encoding both the collagenous (71 bp) as well as the noncollagenous domain [25, 26]. Because the NC1 domain has a primary structure where the first and second halves have a high degree of homology, it has been suggested [4, 29] that the evolution of the NC1 domain involved a duplication of an ancestral coding unit. Analysis of the exon pattern demonstrated, however, that the homology between the two halves of the NC1 domain is not reflected in the exon-intron pattern: there are five exons coding for the NC1 domain, the axis of the symmetry is within one exon, and one of the clusters of amino acids around cysteine residues spans two exons.

Analysis of the 5' end of the gene revealed that the first exon of the α1(IV) gene encodes the 5' untranslated region, the signal peptide and the first residue, lysine, of the secreted form of the α1(IV) chain. The first intron of the α1(IV) gene is large, over 18 kb. The exact size of this intron is not known, since exon 2 was missing in genomic clones isolated, probably because intron 1 contains stretches of repetitive sequences that are unstable in bacteria and are often lost during amplification of libraries. Intron sizes of the α1(IV) gene vary from 86 bp to over 18,000 bp, and repetitive sequences are present in numerous sites in the intervening sequences [28]. Repetitive sequences are often associated with DNA polymorphisms, and in the α1(IV) gene, a polymorphic HindIII site has been detected in the intron 43 [30] in a fragment containing repetitive sequences.

Sequencing of the promoter region of the α1(IV) gene revealed that the gene encoding the α2(IV) chain is very closely located – the transcription start sites of the two genes are only about 100 bp apart both in man and mouse [31–34]. This was the first description of two structural genes in a complex organism having overlapping 5' flanking regions and coding for polypeptide chains of the same protein.

The Gene Encoding α2 Chain of Type IV Collagen

The α2(IV) gene has a size similar to that of the α1(IV) gene [11], but the number of exons is smaller, 47 [35]. In the 5' end of the gene, the first three exons code for the untranslated region and for the signal peptide so that the untranslated region, which is exceptionally long, over 300 bp, is encoded by two exons. The first and second exon are separated by an intron of 330 bp. The second exon contains 44 bp of untranslated sequences and 44 bp coding for the signal peptide. The second intron is again short, 121 bp, and the third exon, 55 bp, encodes part of the signal peptide. The fourth exon contains sequences coding for the three last amino acids of the signal peptide and for the 7S domain. The intron between exons 3 and 4 in the α2(IV) gene is more than 25 kb [28]. Thus the promoter and about 100 bp of coding sequences in both α1(IV) and α2(IV) genes are flanked by intron sequences of 20 kb or more.

Exon sizes in the α1(IV) and α2(IV) collagen genes are different (table 1) but there are similarities between the two genes: In both genes, the exons at the 5' end of the gene start with a complete glycine codon (exons 3–17 in the α2(IV) gene). Beyond that, there are only three exons (22, 34, and the last exon, 47) in the α2(IV) gene that start with a complete codon, all the other

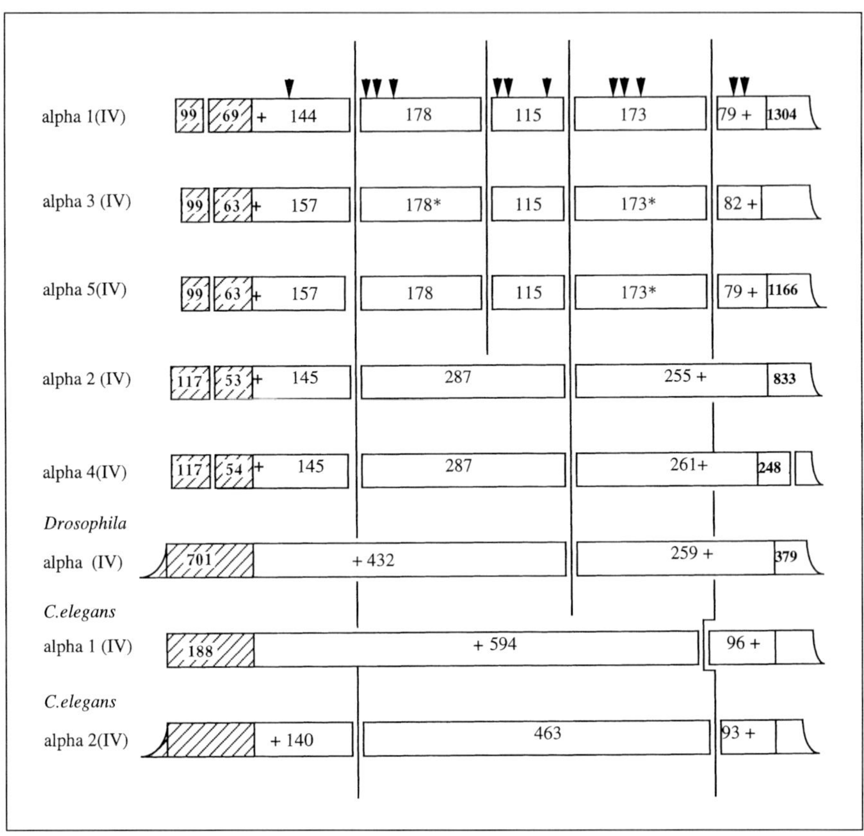

Fig. 3. Exon-intron structure in the NC1-encoding region of α(IV) collagen genes. Exons are depicted by boxes and exon sizes (in bp) are shown inside each box. Intron locations within the NC1 domain coding regions are shown in vertical lines. Collagenous region encoding parts are shaded. Arrowheads indicate locations of cysteine residue-encoding codons. Exons marked with an asterisk are alternatively spliced.

exons start with a split codon [35]. In the α1(IV) gene, exons 2–19, and additionally exons 25, 26, 38, and the second last exon, 51, start with a complete codon [28]. When the two chains are aligned by the similarities in the amino acid sequence, several intron locations match, especially at the aminoterminal region of the collagenous -Gly-X-Y- repeat domain (table 1). Differences are mostly due to differences in the length and locations of interruptions. In the 3' end, the NC1 domain is encoded by three exons in the α2(IV) gene, and the two intron locations have been conserved between the α1(IV) and α2(IV) genes [36] (fig. 3).

The Gene Encoding α5 Chain of Type IV Collagen

The amino acid sequence of the α5(IV) chain shows a high degree of similarity with the α1(IV) chain, 83% in the most conserved NC1 domain [19]. Conservation is visible also in the gene structure: The exon sizes are almost identical with those of the α1(IV) gene especially in the 3' half of the gene (table 1) [37]. Regardless of the conservation of coding sequences, intron sizes do not show any conservation. The gene structure is also very similar in the 5' end of the α1(IV) and α5(IV) genes. In both genes, the first intron is very long and the second exon starts with the last residue of the signal peptide [38]. A feature not seen in the α1 (IV) gene is differential splicing of some exons in the α5(IV) gene. An 18-bp insertion between exons 41 and 42 was reported to be present in kidney mRNA. Within the mRNA, this 18 bp exon is flanked by consensus splice signals [39]. This additional exon codes for two -Gly-X-Y- triplets and does not change the reading frame. The splicing was suggested to be tissue specific since the sequence was not present in mRNA isolated from white blood cells [39]. The significance of this short sequence is not known. The size of the alternatively spliced exon, 18 bp, is unusually short for a collagen gene, but in short-chain collagen genes, especially in the α1(XIII) gene [40], complex splicing patterns including also collagenous exons have been found. Another alternative splicing was found in the 3' end of the α5(IV) gene [41]. An alternatively spliced transcript was detected in human kidney cortex and skin mRNA which skips exon 50 (51 if the 18 bp exon is included) of the α5(IV) collagen gene. This generates a stop codon in the beginning of the last exon, and the result is an α(IV) chain lacking the carboxyterminal half of the NC1 domain. α chains lacking the NC1 domain cannot be incorporated in a triple-helical molecule since the alignment of the chains is started in the NC1 domain [42]. Mutations in the α5(IV) gene have been detected in patients with X chromosome-linked Alport syndrome [43]. (For further description of α5(IV) mutations, see chapter 8.)

The Gene Encoding α6 Chain of Type IV Collagen

The gene encoding α6(IV) chain of type IV collagen was found in a head-to-head arrangement within 450 bp of the α5(IV) gene in X chromosome [22] – an arrangement identical to the α1(IV)-α2(IV) gene organization in chromosome 13, although the distance between the transcription start sites is longer. An interesting difference emerged though, when cDNA clones for the α6(IV) chain were analyzed: In the α6(IV) gene, there are two alternative promoters used in a tissue-type specific fashion [44]. The exon closer to the α5(IV) gene (442 bp away) is 124 bp containing untranslated sequences and 14 bp coding for the first five amino acids of the signal peptide of the protein. The alternative first exon is 246 bp and is located 926 bp downstream (1,368 bp apart from the α5(IV) start site). It contains a longer untranslated segment and 11 bp coding sequences. The second exon is 50 bp encoding part of the signal peptide and is included in both transcripts. Here the α6(IV) gene structure resembles the structure of α2(IV) collagen gene, where the third exon is 55 bp and encodes part of the signal peptide, only the untranslated first exon is missing (see fig. 5). The second intron of the α6(IV) gene is estimated to be over 65 kb in length [45]. The complete exon-intron structure of α6(IV) gene has not been published yet, but the gene has been estimated to be very large, 425 kb [27] – due to extremely long introns.

The Genes Encoding α3 and α4 Chains of Type Collagen

For the α3(IV) and α4(IV) chain genes, only the exons encoding the most 3' end of the gene have been characterized. According to similarities in the amino acid sequence, the chains have been grouped as 'α1-like' and 'α2-like'. This grouping fits to the gene structure as well: the α3(IV) gene structure is similar to the exon-intron pattern of α1(IV) and α5(IV) genes [46] at least in the region coding for the NC1 domain (fig. 3). The α3(IV) chain NC1 domain is encoded by five exons, one of which is a junction exon containing sequences for both the NC1 and the collagenous domain. The 3' border of the α3(IV) chain gene is not known. The intron locations are conserved between the three α1-like genes but intron sizes do not show much similarity [46]. Small differences in the size of the junction exon are due to variation in the length of the NC1 domain in separate α chains. Additionally, length of the 3' untranslated region varies causing differences in the most 3' exon.

Studies on the expression of the α3(IV) mRNA revealed alternative splicing occurring in the 3' end of the α3(IV) RNA, the region encoding the Goodpasture antigen [47, 48]. In the α5(IV) gene, alternative splicing has been detected in the same region [41]. Exons 49 and 51 (numbered according to the α1(IV) gene) can be spliced out either simultaneously or one at a time, hence four different mRNAs are formed. The complete form containing all exon sequences is most common in all tissues studied and expressing α3(IV) but ratio of different splice variants varies in different tissues and during development [47, 48]. Interestingly, alternative splicing was detected in human tissues, but not in bovine or rat. No difference between Goodpasture patients and healthy individuals could be detected [47]. In the shortest variant, all but one cysteine residue are lost from the carboxyterminal end, and in this form the chain cannot be incorporated into a triple-helical molecule. It is not known if individual α chains not integrated into triple-helical structures can compete in cell binding or other functions, or if they are immediately degraded. The most likely function of alternative splicing may be the regulation of amount of mature type IV collagen containing α3(IV) chains – the more splicing into unfunctional forms, the less incorporation into basement membranes in spite of transcriptional activity.

Five exons in the most 3' end of the α4(IV) gene have been characterized from human [49], and two exons from the rabbit gene [14]. Intron locations in this part of the gene parallel intron locations in the α2(IV) gene [36], except that in the α4(IV) gene, there is an additional intron splitting the 3' untranslated region into two exons (fig. 3). The second exon from the 3' end contains 87 complete codons, a termination codon and 245 nucleotides of the 3' untranslated sequence. The rest of the untranslated sequence, about 320 bp, is encoded by a separate exon, an unusual feature in type IV collagen genes.

Invertebrate Type IV Collagen Genes

Collagen IV genes have been characterized in several invertebrates: in the fruit fly *Drosophila* [50], in nematodes *Caenorhabditis elegans* [2, 51] and *Ascaris suum* [52] and in sea urchin, *Strongylocentrotus purpuratus* [53, 54]. Primary structure of the α chains is conserved especially in the carboxyterminal NC1 domain. Additionally, most of the interruptions in the collagenous -Gly-X-Y- sequence in the invertebrate α(IV) chains can be aligned with interruptions in vertebrate α(IV) chains. The gene structures, however, are not conserved.

There is only one type IV collagen gene in *Drosophila* [50]. The gene contains five exons, the sizes of which do not correspond exon sizes in vertebrate collagen genes. Nevertheless, the exon pattern in the most 5' end of the gene is similar to the exon pattern in the vertebrate α2(IV) gene: The first three exons (172, 101 and 49 bp) encode the 5' untranslated region and the signal peptide so that the first exon contains only untranslated sequences, the second exon contains sequences for the untranslated region and for part of the signal peptide, and the short third exon contains signal peptide coding sequences only. The main body of the *Drosophila* α(IV) chain is encoded by long exons of 948, 497, 1452, 979, 1132 and 638 bp. The introns are short, from 62 to 484 bp [50]. In the region coding for the highly conserved NC1 domain, intron 8 is located in the same place as an intron in all vertebrate type IV collagen genes (intron 50 in α1(IV) gene) (fig. 3). Furthermore, there seems to be some sequence conservation in exon sequences bordering intron 6 in the *Drosophila* gene and intron 32 in the human α1(IV) gene [28].

In *C. elegans*, two α(IV) genes have been identified [2]. Even though there is considerable degree of homology in the amino acid sequence in the two chains, there is no conservation of exon sizes or exon-intron boundaries between the two genes [55]. The size of the α1(IV) gene (*clb*-2) is not known, since the 5' and 3' regions of the gene have not been characterized. The analyzed part of the gene, which covers all the translated sequences, is 7 kb in size and contains 12 exons [51]. In the conserved NC1 domain encoding region there is one intron located 14 nucleotides in front of comparable intron location in the mammalian α1(IV) gene (fig. 3). The α2(IV) gene (*clb-1*) is about 9 kb in size and contains 20 exons. Transcription start site is 510 bp upstream from the translation initiation site, and, like in other α2(IV) genes, the most 5' exons are short, and the first exon contains only untranslated sequences. In the NC1 domain coding region there are two introns, one at the identical position with an intron in the vertebrate α1(IV) and α2(IV) genes (intron 48 in the α1(IV) gene), the other one nucleotide past the position of an intron present in the α1(IV) gene and 15 nucleotides in front of an intron in the *C. elegans* α1(IV) gene (fig. 3). Exons 9 (108 bp) and 10 (111 bp) in the *C. elegans* α2(IV) gene, separated by an intron of only 30 bp, are mutually exclusive, and alternative splicing is developmentally regulated so that mRNA containing exon 9 predominates in embryos, while exon 10 containing transcripts are more common in adult RNA. These exons code for similar products differing primarily in the region coding for an interruption in the -Gly-X-Y- sequence [55]. An unusual splice donor sequence was observed at the 5' boundary of intron 10, GCAAG, same sequence as in the 5' end of intron 34 in the human gene [55, 28]. The locations of unusual splice donors do not match between the two genes but still the occurrence of this unusual splice donor in two type IV collagen genes in very distantly related organisms is surprising. Mutations causing embryonic lethality have been characterized in both α1(IV) [51] and α2(IV) [56] genes in *C. elegans*.

Partial structure of the α2(IV) gene of another nematode, *A. suum* has been analyzed from a genomic clone that covers 9 kb and contains 16 exons, 108–490 bp in size. Intron sizes vary from 141 to 854 bp [52]. Since exons of the most 5' and 3' regions of the gene (where exon-intron patterns seem to be most conserved) were not contained in the genomic clone analyzed, comparisons to vertebrate α(IV) cannot be made. Alternative splicing of two exons, 108 and 111 bp, separated by an intron of 39 bp, was discovered coding for the same region as alternatively spliced exons in the *C. elegans* α2(IV) gene. Also in *A. suum* alternative splicing is developmentally regulated [57]. Otherwise the exon sizes in the α2(IV) genes of the two nematodes are not conserved – exons in *A. suum* seem to be shorter than exons in *C. elegans*, especially in the region coding for the carboxyterminal end of the collagenous region. In vertebrates instead, exon sizes are completely conserved between different species.

A genomic clone coding for a collagenous protein called 3α collagen was isolated from sea urchin, *S. purpuratus*, and shown to contain several short exons interrupted by intervening sequences varying in size from 100 to 1,300 bp [53]. Primary structure of 3α collagen is closest to vertebrate α1(IV) chain [54]. 5' untranslated region of 3α reaches 260 bp upstream from the ATG initiation codon and is contained within one exon that also contains sequences for the signal peptide as well as the first codon of the 7S domain, an arrangement identical to the mammalian α1(IV) genes, verifying the classification as α1-like [54]. cDNA clones for the α2(IV) chain of sea urchin have been characterized [58], but a head-to-head arrangement with another gene in the 3α locus was excluded in preliminary characterization of the genomic region [54].

Regulation of Type IV Collagen Genes

The Bidirectional COL4A1-COL4A2 *Promoter*

The genes coding for the type IV collagen α1 and α2 chains are located head-to-head on the same chromosome both in man [33, 34] and mouse [31, 32]. The 5' ends of the genes overlap each other, and the transcription start sites are separated only by approximately 130 bp. The genes are transcribed from opposite DNA strands starting from the short intergenic region which represents the common and bidirectional promoter of α1(IV) and α2(IV) genes. Relative amount of guanosine and cytosine is unusually high in this region which is highly conserved between man and mouse exhibiting nearly 90% sequence homology [31–34]. An incomplete palindromic symmetry in a sequence strech centered by a putative Sp1 binding site (fig. 4) can be distinguished within the bidirectional promoter as well as several inverted repeat streches. Such repeats are often transcription factor binding sites, but they may also have a function in the formation of stable hairpin structures that affect DNA conformation and chromatin structure in this region.

Proximal Regulatory Elements in the COL4A1–COL4A2 *Promoter*

The intergenic promoter region contains several putative binding sites for general transcription factors. There are five GC boxes, binding sites for transcription factor Sp1, one of them located in the center of the common promoter while others are located within 5' regions of both genes (fig. 5). This is a general feature of Sp1-responsive promoters which usually contain multiple binding sites, the most important one being the site closest, approximately 40–70 bp upstream from the transcription start site. Mutational analyses have revealed the importance of the central GC box: Transcriptional activity decreased strikingly in both directions when a mutation was inserted into this sequence [59, 60], while mutations in other GC boxes affected transcriptional activity in much lesser degree. GC boxes are usually functional in both orientations, and GC boxes located in the center of the common promoter and in the first exon of the α2(IV) gene seem to activate both genes. GC boxes in the α1(IV) gene however appear to activate α1(IV) gene only [59]. Therefore,

4

CTC
5' CCGTTAGGCCCCTGTGGGTGCTCCTTGGCCGAGGAGCTCGGTCTTCGCTCTCCCACCCTCCCCCTTTCTACTCCCAAGCAGGAGAGCGTGCAGCCCTAGCCTGCACAAGGCGCTCCAAGT
3' GGCAATCCGGGGACACCCACGAGGAACCGGCTCCTCGAGCCAGAAGCGAGAGGGTGGGAGGGGGAAAGATGAGGGTTCGTCCTCTCGCACGTCGGGATCGGACGTGTTCCGCGAGGTTCA

CGGTGCTCTCGGGCCAGGGTCTGGGCGCCGCTCGGGGGTCGCTCTCACCTCGGGCTCGGCTTCAGGGGCTGCTGCCCGAACGCATTGGCCCTTCCAGAAGCACCCGCCGGCGGCACACCG
GCCACGAGAGCCCGGTCCCAGACCCGCGGCGAGCCCCCAGCGAGAGTGGAGCCCGAGCCGAAGTCCCCGACGACGGGCTTGCGTAACCGGGAAGGTCTTCGTGGGCGGCCGCCGTGTGGC

GC-box   GC-box
GCAGGGCGGCCGGCCTTGCTGGGCTCTCTGGGCGCCGGCCCCGGGGGCTCGGGCGGCCCCTTTCGGTCCTCGGCCTGGATCCGCGAGCCGGCGGGCGCAAGGGGTTGGGACGCGGCAGCC
CGTCCCGCCGGCCGGAACGACCCGAGAGACCCGCGGCCGGGGCCCCCGAGCCCGCCGGGAAAGCCAGGAGCCGGACCTAGGCGCTCGGCCGCCGCGTTCCCCAACCCTGCGCCGTCGG

                                                                  A+T rich
                                                                  element   GC-box
TCTTGAGTGCGGAGCGCGGAGCCCTGGTGTCCCGGCGCACGGCAGCCACACTCCCGGGCCGCGCGCTCCCGCCGCCTCTTACCCGCGCCGCAGGGTCCTCCCCTTTGAGGCGCCGCCCGC
AGAACTCACGCCTCGCGCCTCGGGACCACAGGGCCGCGTGCCGTCGGTGTGAGGGCCCGGCGCGCGAGGGCGGCGGAGAATGGGCGCGGCGTCCCAGGAGGGGAAACTCCGCGGCGGGCG
                             α 2(IV) Exon 1   α 1(IV) Exon 1

CTC   CCAAT
GCACCGCCGGGGGGGAGGGGGCAGCGCCAACAAATTGGGGAGCTCGGCCCGCCGCGCTCAGGTCTCCGCTTGGAGCCGCCGCACCCGGGACGGTGCGTATCGCTGGAAGTCCGGCCTTCC
CGTGGCGGCCCCCCCTCCCCCGTCGCGGTTGTTTAACCCCTCGAGCCGGGCGGCGCGAGTCCAGAGGCGAACCTCGGCGGCGTGGGCCCTGCCACGCATAGCGACCTTCAGGCCGGAAGG
                 A+T rich
                 element

                                                                                                 GC-box
                                                        M G P R L S V W L L L L P A A L L
GAGAGCTAGCTGTCCGCCGCGGCCCCCGCACGCCGGGCAGCCGTCCCTCGCGCCTCGGGCGCGCCACCATGGGGCCCCGGCTCAGCGTCTGGCTGCTGCTGCTGCCCGCCGCCCTTCTGC
CTCTCGATCGACAGGCGGCGCCGGGGGCGTGCGGCCCGTCGGCAGGGAGCGCGGAGCCCGCGCGGTGGTACCCCGGGGCCGAGTCGCAGACCGACGACGACGACGGGCGGCGGGAAGACG

H E E H S R A A A K   GC-box
TCCACGAGGAGCACAGCCGGGCCGCTGCGAAGGTGAGTTCCCGGCCAGCTCCGCTCCCGGCGTCCCGCCCCGAGCTTGGGCGCCCCGAGAGGCCCCTTTGTCCGCGCCTGGACCCGTCGG
AGGTGCTCCTCGTGTCGGCCCGGCGACGCTTCCACTCAAGGGCCGGTCGAGGCGAGGGCCGCAGGGCGGGGCTCGAACCCGCGGGGCTCTCCGGGGAAACAGGCGCGGACCTGGGCAGGC
3'
5'

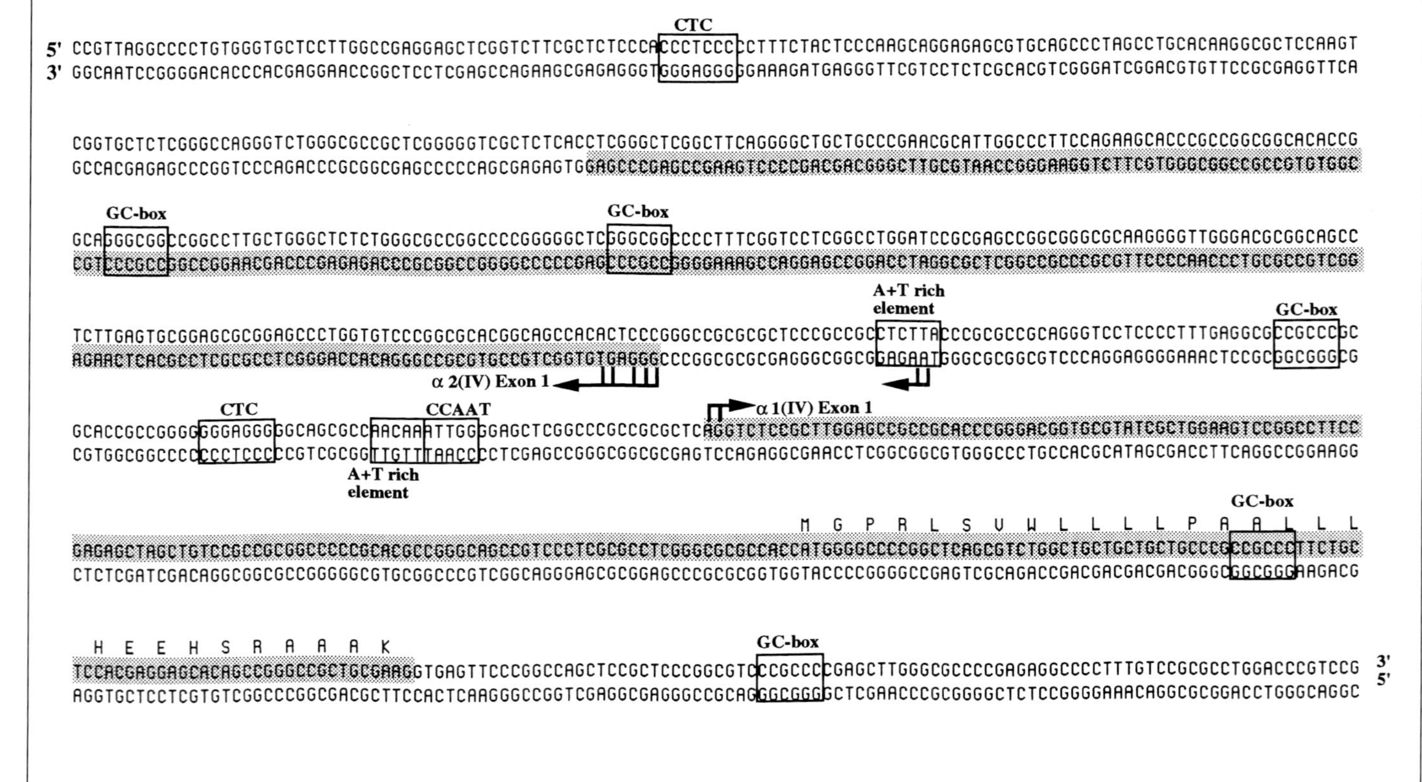

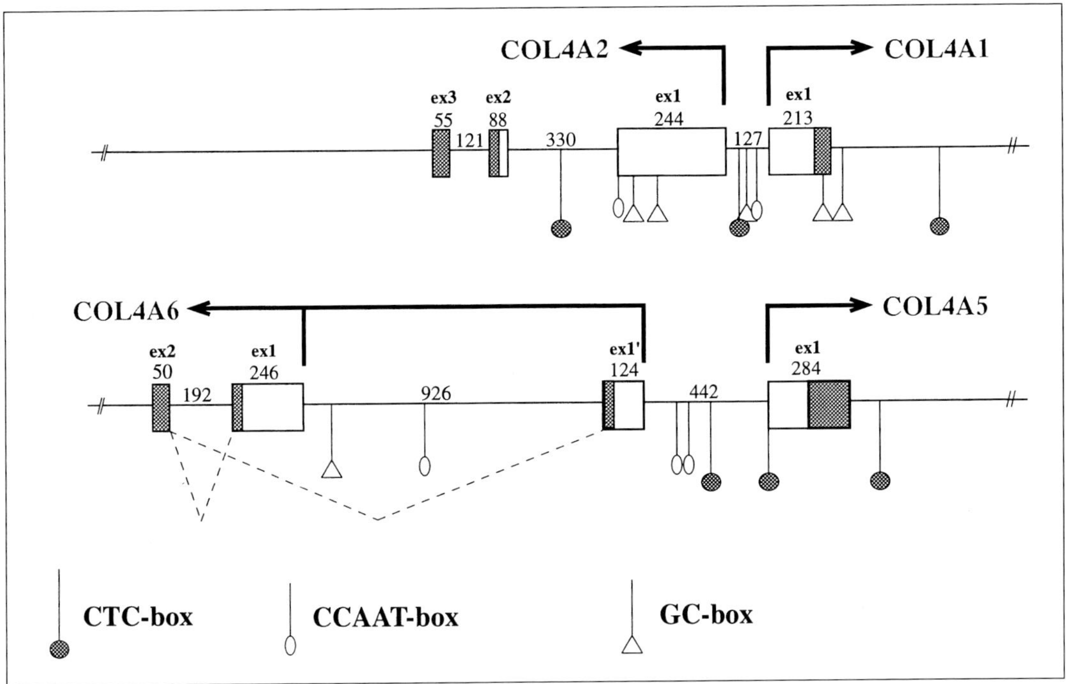

Fig. 5. Promoter region of α1(IV)-α2(IV) (*a*) and A5(IV)-α6(IV) chain genes (*b*). Exons are depicted as boxes, introns as a line. Translated sequences encoding regions are shaded. Sizes of exons and introns are in bp. Transcription factor binding elements are depicted.

GC boxes may have a role in the regulation of the correct ratio of α1(IV) and α2(IV) mRNAs. The region between the two transcription start sites is so short that it is unlikely that two transcription initiation complexes could form simultaneously. Direction of transcription is dictated by factors binding to the promoter region, and a factor bound may cause a steric hindrance which excludes binding of factors favoring the other direction.

The single CCAAT box in the common promoter region has been shown to bind a member of the family of CCAAT binding transcription factors in HT 1080 cells [60]. The CCAAT motif is located 100 bp upstream from the

Fig. 4. Nucleotide sequence of the 5' ends of the human α1(IV) and α2(IV) chain genes. Exon sequences are shaded and amino acids deduced from the exon sequences in the coding regions are depicted by the single letter code. Arrows indicate the transcription initiation sites. Transcription factor binding segments are boxed.

α2(IV) initiation site, and mutations in this site cause a decrease in α2(IV) transcription but not in α1(IV) transcription [59, 60].

A novel regulatory element was characterized in studies on α1(IV)-α2(IV) regulation. A homopyrimidine/purine-rich sequence CCCCCTCCCCCCC (called CTC box) binds transcription factor designated CTCBF, CTC binding factor [61]. CTCBF is composed of two subunits, 75 and 85 kD in size, both of which are needed for sequence-specific interactions with the CTC box [62]. Mutations in the CTC box abolished the binding and decreased transcriptional activity of the promoter in both directions [59–61]. Additionally, another protein with ability to bind the pyrimidine-rich sequence in the bidirectional collagen IV promoter has been cloned [63]. A portion of this protein reveals homology to bacterial DNA ligases as well as to several domains of replication factor C. The role of this protein in collagen IV transcription has not been studied yet.

TATA motifs are AT-rich sequences frequently located 25–30 bp upstream from the transcription initiation site. The consensus sequence TATAA binds transcription factor TF IID, a component of the initiation complex, and is thought to determine the precise location of transcription initiation by RNA polymerase II [64]. The bidirectional α1(IV)-α2(IV) promoter does not contain TATAA consensus sequences. Instead, the promoter contains two short streches rich in AT in positions usually occupied by the TATA motif. By mutational analysis it was shown that disruption of the AT-rich sequence decreased transcription of both genes considerably. When consensus TATAA sequence was inserted into the promoter into the α1(IV) direction, bidirectional transcription was hindered and transcription proceeded mainly to the α1(IV) direction. Instead, when TATA box was inserted into α2(IV) coding strand, α2(IV) transcription was doubled but α1(IV) was not affected [59].

Distal Regulatory Elements in the COL4A1-COL4A2 *Promoter*

In vitro transfection studies with the α1(IV)-α2(IV) promoter constructs containing chloramphenicol acetyltransferase (CAT) reporter gene in either α1 or α2 direction have shown that the promoter region alone has very low activity [33]. Therefore, regulatory elements located within the genes must control the cell type-specific expression. A distal enhancer element located in the first intron of the murine α1(IV) gene has been shown to have high activity in both the α1(IV) and α2(IV) direction and also with a heterologous promoter [31, 65]. Transient transfection studies delineated the enhancer to a 210-bp fragment located about 2.7 kb downstream from the α1(IV) promoter. Two *cis*-acting elements were identified in this region, and one of the elements containing the sequence CCTTATCTCTGATGG motif appears to be a novel nuclear factor binding site for two putative transcription factors. Mutations in this motif abolished factor binding and caused a significant decrease in enhancer activity [66]. Furthermore, another cell type-specific enhancer was identified [67] in close vicinity to the 210-bp enhancer described above. This enhancer was delimited to a 300-bp fragment found 4.5 kb upstream from the $\alpha2(IV)$ collagen gene. It confers a cell type-specific activation on the $\alpha1(IV)$-$\alpha2(IV)$ promoter and on a heterologous promoter as well. Of interest is that the first nine nucleotides of the CCTTATCTCTTGATGG enhancer motif [66] are also found in the 300-bp enhancer fragment. The two short regions together may form a cell-specific enhancer regulating expression of the $\alpha1(IV)$-$\alpha2(IV)$ gene locus. In the human *COL4A1-COL4A2* locus, enhancers have not been found in corresponding positions [59]. Introns in this region are large and contain repetitive sequences, and regulatory elements in the human genes may be embedded in the intron sequences further downstream.

A silencing element has been suggested to be located in the third intron of human $\alpha2(IV)$ gene [70]. This element is able to inhibit transcription of a heterologous gene and both type IV collagen genes, and seems to be independent of the distance to the promoter or its direction.

Transcription in the $\alpha2(IV)$ direction is strictly dependent on the intact genomic structure of the region, and it has been suggested [68] that an activating element is located within the first intron of the $\alpha2(IV)$ gene. On the other hand, it has been shown [69] that intervening sequences in the 5' region of a gene increase expression considerably. As an intron has been found splitting the long 5' untranslated region in all $\alpha2(IV)$ genes analyzed, it may be that the intron is needed for correct processing of mRNA rather than activating the transcription.

In general, it must be concluded that the regulation of $\alpha1(IV)$-$\alpha2(IV)$ gene expression is not well understood in spite of extensive studies. The ratio of $\alpha1(IV)$ and $\alpha2(IV)$ mRNAs varies in different cells and conditions, but the ratio of α chains in the triple-helical molecule is constant. Consequently, expression must be extensively regulated on different levels: on the level of transcription by regulating transcription initiation and progress (some elements playing a role in these processes are described above), but also on regulating stability and processing of RNA, as well as efficiency of translation, posttranslational modification of α chains, and incorporation of triple-helical molecules in basement membranes.

The Bidirectional COL4A5-COL4A6 *Promoter*

The $\alpha5(IV)$ and $\alpha6(IV)$ genes are located head-to-head on chromosome X, a situation identical to the arrangement of $\alpha1(IV)$-$\alpha2(IV)$ genes [22]. However, the distance between transcription start sites is longer, 442 bp (fig. 5). Furthermore, the $\alpha6(IV)$ gene has two alternative promoters, so that the further initiation site is 1,368 bp away from the α5(IV) transcription start site [44]. The existence of two alternative start sites indicates usage of different regulatory elements for the expression of the α6(IV) gene. Transcripts containing the more upstream first exon (1') are abundant in placenta whereas exon 1 containing transcripts are more frequent in kidney and lung [44]. It can be hypothesized that exon 1' is used as a starting site in tissues where both α5(IV) and α6(IV) genes are expressed, and regulation is governed by elements located in the α5(IV) gene. In contrast, the longer α6(IV) promoter is used when only the α6(IV) is expressed and activating elements are located either in close vicinity to exon 1 or further downstream.

Significant homology between the α1(IV)-α2(IV) and α5(IV)-α6(IV) promoter regions cannot be detected. Enrichment of guanosine and cytosine is seen especially upstream from the α5(IV) gene and upstream exon 1 of α6(IV) gene, stretches of TA are more common in the exon 1' and between exons 1' and 1 in the α6(IV) gene.

A cryptic promoter was suggested to be located in the long second intron of the α6(IV) since a truncated α6(IV) mRNA was detected in patients having deletions in the 5' end of α6(IV) gene [45]. Regulatory elements located in the intron are probably used in initiation from the cryptic initiation site.

Evolution of Type IV Collagen Genes

The structure of the genes coding for fibrillar collagens is highly homologous: equivalent amino acid segments in the triple-helical domain are encoded by exons of equal size although the sequences in these exons are different. Most of the exons coding for the triple-helical domain are of 54 bp or multiples of this size. It has been proposed [71] that the genes for fibrillar collagens evolved by amplification of a single genetic unit of 54 bp embedded in intron sequences. The gene structure of α(IV) chain genes is distinctly different from the structure of fibrillar collagen genes, but some features may reflect the common origin: In the vertebrate α1(IV) genes, there are two exons of 54 bp coding for the -Gly-X-Y- repeat sequence, in the α2(IV) gene one exon of 54 bp and two exons of 108 bp. In fibrillar collagen genes, all exons start with a complete codon, in α1(IV) and α2(IV) genes, exons coding for the amino-terminal region of collagenous domain start with a complete codon, other exons with a split codon. The rigid gene structure of fibrillar collagen genes is presumed to reflect the requirement to maintain an exactly equal length of α chains for a correct assembly of collagen molecules into fibrils [71]. The α chains of fibrillar collagens have continuous collagenous domains whereas α chains of type IV collagen have frequent interruptions in the -Gly-X-Y-

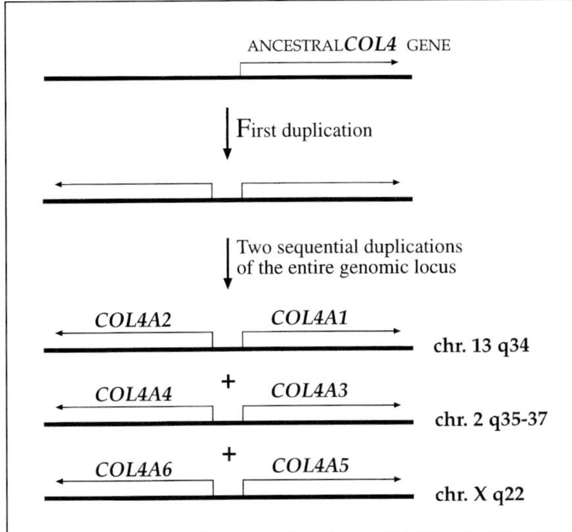

Fig. 6. Evolution of type IV collagen genes.

repeat sequence in the collagenous domain. Therefore, changes in the exon pattern of the collagenous region in α(IV) genes have been allowed during evolution.

Analysis of the occurrence of interruptions in the exons in the α2(IV) gene revealed that interruptions are often located at the intron-exon borders, which could mean that they were recruited from intron sequences. This could happen if the splice site is inactivated by a mutation and a cryptic splice site nearby is taken in use. Interruptions within exons could have evolved from intron losses between exons that had earlier taken up intron sequences, or from incomplete losses of intron sequences during a fusion of two exons [35].

The genes for the α1(IV) and α2(IV) and other closely related α chains of type IV collagen are presumed to have evolved from a common ancestor that was duplicated and inverted so that genes were arranged head-to-head. The genomic region containing the two genes was then further duplicated, and the end result is three pairs of closely related type IV collagen genes in vertebrates (fig. 6). Interruption sequences are totally unrelated between α1(IV) and α2(IV) genes, and it has been suggested [35] that the ancestor α(IV) collagen gene was expressed without interruptions. Interruptions evolved then independently after the gene duplication by recruitment of intron sequences and were selected by position and approximate length. The first duplication as well as recruitment of some interruptions have happened at least 700 million years ago [36], before the separation of vertebrate and invertebrate lines, since there are 14 interruptions in the *Drosophila* α(IV) gene, of which all but one have equivalents in either or both of vertebrate α1(IV) and α2(IV) chains [50], 15 interruptions in the α2(IV) and 17 interruptions in the α1(IV) chain in *C. elegans* [51, 55]. *Drosophila* is supposed to have lost one of the α chains during evolution, and its α(IV) chain may be a combination of early α1(IV) and α2(IV) chains.

Vertebrate α(IV) collagen genes are composed of numerous short exons separated by long introns, invertebrate genes are much shorter containing long exons and short introns. Introns in collagen genes are supposed to be important in impeding recombinations between repetitive exons [72] or in stabilizing the RNA structure, and ensuring correct splicing. Some introns may indeed have this kind of function but they may also reflect the ancient assembly of genes from small units, which has been retained in vertebrates but lost in invertebrates to achieve more rapid replication of the genome.

The primary structure of the NC1 domain is well conserved reflecting its importance in type IV collagen assembly. Its composition of two homologous domains suggests duplication of an ancestral exon that encoded sequences for about 115 amino acids, but the structure of α(IV) genes reveals a more complicated evolution comprising various changes in the exon pattern. Possibly each original cluster around cysteines was encoded by a separate exon, the region containing three exons was duplicated, and some of the introns were deleted during the development. Introns conserved during evolution of different species often have important functions in gene regulation. The intron in the NC1 domain coding region of α(IV) collagen genes may have a regulatory function but may as well be needed for stability of RNA because of repetitive nature of the domain. Alternative splicing seen in this region in the α3(IV) and α5(IV) chain RNAs may reflect instability in this part of the gene. Similarly, the intron in the most 5' end of α2(IV) genes might either contain binding elements for regulatory proteins, or have a function in ensuring correct splicing of RNA. It has been suggested that a splicing machinery utilizes one or both ends of the primary transcript as an initial reference point for the selection of donor and acceptor junctions in constructing mRNAs [73].

The significance of the close linkage of two α(IV) genes in a head-to-head arrangement in vertebrates is not known. There has to be selection to maintain the genes in a cluster after duplication. It is supposed that a tight linkage of closely related genes would result in an unsupportable amount of recombination or gene conversion, and separation of genes is favored because new developmental or regulatory functions are created – as has happened to the α(IV) genes during invertebrate evolution. Close linkage of genes favors coordinated transcription which can be controlled by joint regulatory factors. This may be especially important during early development of vertebrates. Furthermore, regulatory elements needed for expression are probably located within the other gene of the pair, and development of similar elements separately might require so much time that it has not happened yet at least completely. Promoter structure of the α3(IV)-α4(IV) unit has not been analyzed yet. It will be interesting to see if it is a similar close package as α1(IV)-α2(IV) promoter or a wider arrangement like the α5(IV)-α6(IV) unit. At present, two possibilities can be envisaged: Either the ancestor unit was tightly controlled by intermingled regulatory elements, and α5(IV)-α6(IV) promoter has developed to a more relaxed arrangement so that α6(IV) chain is expressed also independently. Or, in the original unit genes were separated, and evolution of α1(IV)-α2(IV) genes has led to the very close arrangement seen today.

New functions of α(IV) chains have evolved by duplication of the gene pair unit. α3(IV)-α4(IV) unit has evolved further than the other two gene pairs, which indicates that the α3(IV)-α4(IV) duplicate was first to separate, and α1(IV)-α2(IV) and α5(IV)-α6(IV) are later duplicates. An interesting feature is the chromosomal localization of α(IV) gene pairs: both the *COL4A3-COL4A4* as well as the *COL4A5-COL4A6* pair are located in chromosomal regions that are well conserved between man and mouse, instead, genes in the terminal long arm of human chromosome 13, near to the *COL4A1-COL4A2* region, are more dispersed to different chromosomes in mouse [74]. It could be speculated that this, somehow unstable region contained the ancestor α(IV) gene – therefore it was easily duplicated and duplicates were then stably integrated in different regions of the genome. No evidence to support this hypothesis is available yet but new information about genome organization is rapidly accumulating from genome sequencing and comparative mapping of human and mouse genomes, which will provide more information about evolution of genes and genomes.

Analysis of type IV collagen genes has supplied new information not only about basement membrane structure, function, and diseases connected to it, but also about gene evolution and regulation. Further studies especially on regulation of type IV collagen genes will be needed in order to better understand diseases linked to these genes, and possibly also for applications in gene therapy.

References

1 Blumberg B, MacKrell AJ, Olson PF, Kurkinen M, Monson JM, Natzle JE, Fessler JH: Basement membrane procollagen IV and its specialized carboxyl domain are conserved in *Drosophila*, mouse, and human. J Biol Chem 1987;262:5947–5950.

2 Guo X, Kramer JM: The two *Caenorhabditis elegans* basement membrane (type IV) collagen genes are located on separate chromosomes. J Biol Chem 1989;264:17574–17582.

3 Oohashi T, Sugimoto M, Mattei M-G, Ninomiya Y: Identification of a new collagen IV chain α6(IV), by cDNA isolation and assignment of the gene to chromosome Xq22, which is the same locus for COL4A5. J Biol Chem 1994;269:7520–7526.

4 Pihlajaniemi T, Tryggvason K, Myers JC, Kurkinen M, Lebo R, Cheung M-C, Prockop DJ, Boyd CD: cDNA clones coding for the pro-α1(IV) chain of human type IV procollagen reveal an unusual homology of amino acid sequences in two halves of the carboxyl-terminal domain. J Biol Chem 1985;260:7681–7687.

5 Solomon E, Hiorns LR, Spurr N, Kurkinen M, Barlow D, Hogan BLM, Dalgleish R: Chromosomal assignments of the genes coding for human types II, III, and IV collagen: A dispersed gene family. Proc Natl Acad Sci USA 1985;82:3330–3334.

6 Boyd CD, Weliky K, Toth-Fejel SE, Deak SB, Christiano AM, Mackenzie JW, Sandell LJ, Tryggvason K, Magenis E: The single copy gene coding for human α1(IV) procollagen is located at the terminal end of the long arm of chromosome 13. Hum Genet 1986;74:121–125.

7 Emanuel BS, Sellinger BT, Gudas LJ, Myers JC: Localization of the human procollagen α1(IV) gene to chromosome 13q34 by in situ hybridization. Am J Hum Genet 1986;38:38–44.

8 Griffin CA, Emanuel BS, Hansen JR, Cavenee WK, Myers JC: Human collagen genes encoding basement membrane α1(IV) and α2(IV) chains map to the distal long arm of chromosome 13. Proc Natl Acad Sci USA 1987;84:512–516.

9 Killen PD, Francomano CA, Yamada Y, Modi WS, O'Brien SJ: Partial structure of the human α2(IV) collagen chain and chromosomal localization of the gene (COL 4A2). Hum Genet 1987;77: 318–324.

10 Boyd CD, Toth-Fejel S, Gadi IK, Litt M, Condon MR, Kolbe M, Hagen IK, Kurkinen M, Mackenzie JW, Magenis E: The genes coding for human pro alpha 1(IV) and pro alpha 2(IV) collagen are both located at the end of the long arm of chromosome 13. Am J Hum Genet 1988;42:309–314.

11 Cutting GR, Kazazian HH Jr, Antonarakis SE, Killen PD, Yamada Y, Francomano CA: Macrorestriction mapping of COL4A1 and COL4A4 collagen genes on human chromosome 13q34. Genomics 1988;3:256–263.

12 Bowcock AM, Hebert JM, Wijüsman E, Gadi I, Cavalli-Sforza LL, Boyd CD: High recombination between two physically close human basement membrane collagen genes at the distal end of chromosome 13q. Proc Natl Acad Sci USA 1988;85:2701–2705.

13 Morrison KE, Mariyama M, Yang-Feng TL, Reeders ST: Sequence and localization of a partial cDNA encoding the human α3 chain of type IV collagen. Am J Hum Genet 1991;49:545–554.

14 Kamagata Y, Mattei M-G, Ninomiya Y: Isolation and sequencing of cDNAs and genomic DNAs encoding the α4 chain of basement membrane collagen type IV and assignment of the gene to the distal long arm of human chromosome 2. J Biol Chem 1992;267:23753–23758.

15 Mariyama M, Zheng K, Yang-Feng TL, Reeders ST: Colocalization of the genes for the α3(IV) and α4(IV) chains of type IV collagen to chromosome 2 bands q35–q37. Genomics 1992;13:809–813.

16 Weil D, Mattei M-G, Passage E, Vang Cong N, Pribula-Conway D, Mann K, Deutzmann R, Timpl R, Chu M-L: Cloning and chromosomal localization of human genes encoding the three chains of type VI collagen. Am J Hum Genet 1988;42:435–445.

17 Emanuel BS, Cannizzaro LA, Seyer JM, Myers JC: Human α1(III) and α2(V) procollagen genes are located on the long arm of chromosome 2. Proc Natl Acad Sci USA 1985;82:3385–3389.

18 Schurr E, Skamene E, Morgan K, Chu M-L, Gros P: Mapping of Col3a1 and Col6a3 to proximal murine chromosome 1 identifies conserved linkage of structural protein genes between murine chromosome 1 and human chromosome 2q. Genomics 1990;8:477–486.

19 Hostikka SL, Eddy RL, Byers MG, Höyhtyä M, Shows TB, Tryggvason K: Identification of a distinct type IV collagen α chain with restricted kidney distribution and assignment of the gene to the locus of X chromosome-linked Alport syndrome. Proc Natl Acad Sci USA 1990;87:1606–1610.

20 Myers JC, Jones TA, Pohjolainen E-R, Kadri AS, Goddard AD, Sheer D, Solomon E, Pihlajaniemi T: Molecular cloning of α5(IV) collagen and assignment of the gene to the region of the X chromosome containing the Alport syndrome locus. Am J Hum Genet 1990;46:1024–1033.

21 Brown SDM, Avner P, Boyd Y, Chapman V, Rastan S, Sefton L, Thomas JD, Herman GE: Mouse X chromosome. Mammalian Genome 1993;4:S269–S281.

22 Zhou J, Mochizuki T, Smeets H, Antignac C, Laurila P, de Paepe A, Tryggvason K, Reeders ST: Deletion of the paired α5(IV) and α6(IV) collagaen genes in inherited smooth muscle tumors. Science 1993;261:1167–1169.

23 Kurkinen M, Bernard MP, Barlow DP, Chow LT: Characterization of 64-, 123-, and 182-base pair exons in the mouse α2(IV) collagen gene. Nature 1985;317:177–179.

24 Sakurai Y, Sullivan M, Yamada Y: α1 type IV collagen gene evolved differently from fibrillar collagen genes. J Biol Chem 1986;261:6654–6657.

25 Soininen R, Chow L, Kurkinen M, Tryggvason K, Prockop DJ: The gene for the α1(IV) chain of human type IV procollagen: The exon structures do not coincide with the structural subdomains in the globular carboxy-terminus of the protein. EMBO J 1986;5:2821–2823.

26 Soininen R, Tikka L, Chow L, Pihlajaniemi T, Kurkinen M, Prockop DJ, Boyd CD, Tryggvason K: Large introns in the 3'-end of the gene for the proα1(IV) chain of human basement membrane collagen. Proc Natl Acad Sci USA 1986;83:1568–1572.

27 Srivastava AK, Featherstone T, Wein K, Schlessinger D: YAC contigs mapping the human COL4A5 and COL4A6 genes and DXS118 within Xq21.3–q22. Genomics 1995;26:502–509.

28 Soininen R, Huotari M, Ganguly A, Prockop DJ, Taryggvason K: Structural organization of the gene for the α1(IV) chain of human type IV collagen. J Biol Chem 1989;264:13565–13571.

29 Brinker JM, Gudas LJ, Loidl HR, Wang S-Y, Rosenbloom J, Kefalides NA, Myers JC: Restricted homology between human α1 type IV and other procollagen chains. Proc Natl Acad Sci USA 1985; 82:3649–3653.

30 Tikka L, Roiko K, Soininen R, Prockop DJ, Tryggvason K: A HindIII polymorphism in the 3' end of the human α1(IV) collagen gene. Nucleic Acids Res 1987;15:5497.

31 Burbelo PD, Martin GR, Yamada Y: α1(IV) and α2(IV) collagen genes are regulated by a bidirectional promoter and a shared enhancer. Proc Natl Acad Sci USA 1988;85:9679–9682.

32 Kaytes P, Wood L, Theriault N, Kurkinen M, Vogeli G: Head-to-head arrangement of murine type IV collagen genes. J Biol Chem 1988;263:19274–19277.

33 Pöschl E, Pollner R, Kühn K: The genes for the α1(IV) and α2(IV) chains of human basement membrane collagen type IV are arranged head-to-head and separated by a bidirectional promoter of unique structure. EMBO J 1988;7:2687–2695.

34 Soininen R, Huotari M, Hostikka SL, Prockop DJ, Tryggvason K: The structural genes for α1 and α2 chains of human type IV collagen are divergently encoded on opposite DNA strands and have an overlapping promoter region. J Biol Chem 1988;263:17217–17220.

35 Buttice G, Kaytes P, D'Armiento J, Vogeli G, Kurkinen M: Evolution of collagen IV genes from a 54-base pair exon: A role for introns in gene evolution. J Mol Evol 1990;30:479–488.

36 Hostikka SL, Tryggvason K: Extensive structural differences between genes for the α1 and α2 chains of type IV collagen despite conservation of coding sequences. FEBS Lett 1987;224:297–305.

37 Zhou J, Hostikka SL, Chow LT, Tryggvason K: Characterization of the 3' half of the human type IV collagen A5 gene that is affected in Alport syndrome. Genomics 1991;9:1–9.

38 Zhou J, Leinonen A, Tryggvason K: Structure of the human type IV collagen *COL4A5* gene. J Biol Chem 1994;269:6608–6614.

39 Guo C, van Damme B, van Damme-Lompaerts R, van den Berghe H, Cassiman J-J, Marynen P: Differential splicing of COL4A5 mRNA in kidney and white blood cells: A complex mutation in the COL4A5 gene of an Alport patient deletes the NC1 domain. Kidney Int 1993;44:1316–1321.

40 Tikka L, Pihlajaniemi T, Henttu P, Prockop DJ, Tryggvason K: Gene structure for the α1 chain of human short chain collagen (type XIII) with alternatively spliced transcripts and translation termination codon at the 5' end of the last exon. Proc Natl Acad Sci USA 1988;85:7491–7495.

41 Saito A, Sakatsume M, Yamazaki H, Arakawa M: Alternative splicing in the alpha5(IV) collagen gene in human kidney and skin tissues. Nippon Jinzo Gakki Shi 1994;36:19–24.

42 Dölz R, Engel J, Kühn K: Folding of collagen IV. Eur J Biochem 1988;178:357–366.

43 Barker DF, Hostikka SL, Zhou J, Chow LT, Oliphant AR, Gerken SC, Gregory MC, Skolnick MH, Atkin CL, Tryggvason K: Identification of mutations in the COL4A5 collagen gene in Alport syndrome. Science 1990;248:1224–1227.

44 Sugimoto M, Oohashi T, Ninomiya Y: The genes COL4A5 and COL4A6, coding for basement membrane collagen chains α5(IV) and α6(IV), are located head-to-head in close proximity on human chromosome Xq22 and COL4A6 is transcribed from two alternative promoters. Proc Natl Acad Sci USA 1994;91:11679–11683.

45 Heidel L, Dahan K, Zhou J, Zhang X, Cochat P, Gould JDM, Leppig KA, Proesmans W, Guyot C, Guillot M, Roussel B, Tryggvason K, Grünfeld J-P, Gubler M-C, Antignac C: Deletions of both α5(IV) and α6(IV) collagen genes in Alport syndrome and in Alport syndrome associated with smooth muscle tumours. Hum Mol Genet 1995;4:99–108.

46 Quinones S, Bernal D, Carcia-Sogo M, Elena SF, Saus J: Exon/intron structure of the human α3(IV) gene encompassing the Goodpasture antigen (α3(IV)NC1). Identification of a potentially antigenic region at the triple helix/NC1 domain junction. J Biol Chem 1992;267:19780–19784.

47 Bernal D, Quinones S, Saus J: The human mRNA encoding the Goodpasture antigen is alternatively spliced. J Biol Chem 1993;268:12090–12094.

48 Feng L, Xia Y, Wilson CB: Alternative splicing of the NC1 domain of the human α3(IV) collagen gene. Differential expression of mRNA transcripts that predict three protein variants with distinct carboxyl regions. J Biol Chem 1994;269:2342–2348.

49 Sugimoto M, Oohashi T, Yoshioka H, Matsuo N, Ninomiya Y: cDNA isolation and partial gene structure of the human α4(IV) collagen chain. FEBS Lett 1993;330:122–128.

50 Blumberg B, MacKrell AJ, Fessler JH: Drosophila basement membrane procollagen α1(IV). II. Complete cDNA sequence, genomic structure and general implications for supramolecular assemblies. J Biol Chem 1988;263:18328–18337.

51 Guo X, Johnson JJ, Kramer JM: Embryonic lethality caused by mutations in basement membrane collagen of *C. elegans*. Nature 1991;349:707–709.

52 Pettitt J, Kingston IB: The complete primary structure of a nematode α2(IV) collagen and the partial structural organization of its gene. J Biol Chem 1991;266:16149–16156.

53 Venkatesan M, de Pablo F, Vogeli G, Simpson RT: Structure and developmentally regulated expression of a *Strongylocentrotus purpuratus* collagen. Proc Natl Acad Sci USA 1986;83:3351–3355.

54 Exposito JY, D'Alessio M, Di Liberto M, Ramirez F: Complete primary structure of a sea urchin type IV collagen α chain and analysis of the 5' end of its gene. J Biol Chem 1993;268:5249–5254.

55 Sibley MH, Johnson JJ, Mello CC, Kramer JM: Genetic identification, sequence, and alternative splicing of the *Caenorhabditis elegans* α2(IV) collagen gene. J Cell Biol 1993;123:255–264.

56 Sibley MH, Graham PL, von Mende N, Kramer JM: Mutations in the α2(IV) basement membrane collagen gene of *Caenorhabditis elegans* produce phenotypes of differing severities. EMBO J 1994;13:3278–3285.

57 Pettitt J, Kingston IB: Developmentally regulated alternative splicing of a nematode type IV collagen gene. Dev Biol 1994;161:22–29.

58 Exposito JY, Suzuki H, Geourjon C, Garrone R, Solursh M, Ramirez F: Identification of a cell lineage-specific gene coding for a sea urchin α2(IV)-like collagen chain. J Biol Chem 1994;269:13167–13171.

59 Heikkilä, Soininen R, Tryggvason K: Directional regulatory activity of *cis*-acting elements in the bidirectional α1(IV) and α2(IV) collagen gene promoter. J Biol Chem 1993;268:24677–24682.

60 Schmidt C, Fischer G, Kadner H, Gensersch E, Kühn K, Pöschl E: Differrential effects of DNA-binding proteins on bidirectional transcription from the common promoter region of human collagen type IV genes COL4A1 and COL4A2. Biochim Biophys Acta 1993;1174:1–10.

61 Fischer G, Schmidt C, Opitz J, Cully Z, Kühn K, Pöschl E: Identification of a novel sequence element in the common promoter region of human collagen type IV genes, involved in the regulation of divergent transcription. Biochem J 1993;292:687–695.

62 Genersch E, Eckerskorn C, Lottspeich F, Herzog C, Kühn K, Pöschl E: Purification of the sequence-specific transcription factor CTCBF, involved in the control of human collagen IV genes: Subunits with homology to Ku antigen. EMBO J 1995;14:791–800.

63 Burbelo PD, Utani A, Pan Z-Q, Yamada Y: Cloning of the large subunit of activator 1 (replication factor C) reveals homology with bacterial DNA ligases. Proc Natl Acad Sci USA 1993;90:11543–11547.

64 Sawadogo M, Sentenac A: RNA polymerase B (II) and general transcription factors. Annu Rev Biochem 1990;59:711–754.

65 Killen PD, Burbelo PD, Martin GR, Yamada Y: Characterization of the promoter for the α1(IV) collagen gene. DNA sequencing within the first intron enhance transcription. J Biol Chem 1988;263:12310–12314.

66 Burbelo PD, Bruggeman LA, Gabriel GC, Klotman PE, Yamada Y: Characterization of a cis-acting element required for efficient transcriptional activation of the collagen IV enhancer. J Biol Chem 1991;266:22297–22302.

67 Tanaka S, Kaytes P, Kurkinen M: An enhancer for transcription of collagen IV genes is activated by F9 cell differentiation. J Biol Chem 1993;268:8862–8870.

68 Pollner R, Fischer G, Pöschl E, Kühn K: Regulation of divergent transcription of the genes coding for basement membrane type IV collagen. Ann NY Acad Sci 1990;580:44–54.

69 Buchman AR, Berg P: Comparison of intron-dependent and intron-independent gene expression. Mol Cell Biol 1988;8:4395–4405.

70 Yamada Y, Kühn K: Genes and regulation of basement membrane collagen and laminin synthesis; in Rohrbach DH, Timpl R (eds): Molecular and Cellular Aspects of Basement Membranes. San Diego, Academic Press, 1993, pp 121–146.

71 Yamada Y, Avvedimento VE, Mudryj M, Ohkubo H, Vogeli G, Irani M, Pastan I, de Crombrugghe B: The collagen gene: Evidence for its evolutionary assembly by amplification of a DNA segment containing an exon of 54 bp. Cell 1980;22:887–892.

72 Ramirez F, Bernard M, Chu M-L, Dickson L, Sangiorgi F, Weil D, de Wet W, Junien C, Sobel M: Isolation and characterization of the human fibrillar collagen genes. Ann NY Acad Sci 1985;460:117–124.

73 Wang B, Korfhagen TR, Gallagher PM, D'Amore M, McNeish J, Potter SS, Ganschow RE: Overlapping transcriptional units on the same strand within the murine β-glucuronidase gene complex. J Biol Chem 1988;263:15841–15844.

74 Lyon MF, Kirby MC: Mouse chromosome atlas. Mouse Genome 1995;93:23–66.

Raija Soininen, PhD, Biocenter Oulu and Department of Biochemistry, University of Oulu, FIN–90570 Oulu (Finland)

Tryggvason K (ed): Molecular Pathology and Genetics of Alport Syndrome.
Contrib Nephrol. Basel, Karger, 1996, vol 117, pp 130–141

Distribution of Type IV Collagen α1, α2 and α5 Chains in Human Tissues

Hannu Sariola[a], *Sirkka Liisa Hostikka*[b], *Sari Lukkarila*[b],
Karl Tryggvason[b]

[a] Institute of Biotechnology and Department of Pathology, University of Helsinki,
and [b] Biocenter Oulu and Department of Biochemistry, University of Oulu, Finland

Type IV collagen, the major basement membrane collagen, is composed of three α chains that form a triple helical molecule [1]. Each α chain is approximately 400 nm long and consists of a long aminoterminal collagenous domain with several imperfections in the Gly-X-Y-repeat sequence and a carboxylterminal globular noncollagenous (NC1) domain. The most common form of type IV collagen consists of two α1(IV) and one α2(IV) chains. Four additional distinct type IV collagen α chains, α3(IV), α4(IV), α5(IV) and α6(IV), have been characterized [2–9]. These chains possess a striking tissue specificity [3, 8–12]. The α3(IV) and α4(IV) chains are probably present in the same molecule [13] but it is not known with which chains α5(IV) can combine.

Using a monoclonal antibody to the NC domain of the human α1(IV) chain, Kleppel et al. [10] have shown that this chain is located in most, if not all, basement membranes. A similar study with α2(IV) chain-specific antibodies has not been performed, although it can be anticipated to have the same distribution, since it is usually present together with the α1(IV) chain in a heterotrimer molecule. Antibodies raised against isolated 28 kD NC domains M28+++ and M28+ derived from the α3(IV) and α4(IV) chains, respectively, have shown a clearly more limited, but identical distribution of these two chains [10, 11]. For example, in kidney cortex antibodies to both chains stained strongly the glomerular basement membrane (GBM), the distal tubular basement membrane (TBM) and the Bowman's capsule. They were shown to stain strongly the alveolar, but not capillary basement membranes (BMs) of lung. Both chains are present in the choroid plexus ependymal BM, but to a lesser extent in choroid vessels and gray and white matter BMs. Furthermore, the α3(IV) and α4(IV) chains showed high region specificity in the cochlea and eye [12]. Sanes et al. [14] have colocalized the α3(IV) and α4(IV) chains in the BMs of synaptic muscle fibers. Interestingly, the α3(IV) and α4(IV) chains are absent from the epidermal BM as well as the BMs of placenta, spleen, liver, skeletal muscle, arteries and endoneural and perineural nerve [11, 14]. These findings have gained support from other studies [12, 15].

The α5(IV) chain whose gene has been shown to be mutated in X-linked Alport syndrome [16–18] has been localized in kidney cortex solely to the GBM [3]. However, based on Northern analyses it is known to be expressed in lung and spleen as well [3]. The more limited distribution of the α5(IV) than that of the α3(IV) and α4(IV) chains in kidney indicates that they may not always be present in the same triple helical molecule.

The NC1 domain of the α3(IV) chain has been identified in several studies [2, 19–23] as the main target for autoimmune antibodies in Goodpasture syndrome, a disease characterized by lung hemorrhage and glomerulonephritis. Interestingly, male patients with Alport syndrome have been reported to lack peptides corresponding to the α3(IV) and α4(IV) chains and, similarly, Goodpasture antisera have been shown not to stain the GBM of Alport patients [24]. Alport syndrome is characterized by hematuria, hearing loss and terminal renal failure, but it lacks the lung symptoms characteristic for the Goodpasture syndrome. The association of the α3(IV) and α4(IV) collagen chains and their genes, which are located on chromosome 2 [25], with X-linked Alport syndrome is still not understood.

To further elucidate the tissue distribution of type IV collagen subunits, we have made antibodies against synthetic oligopeptides and used them in an immunohistochemical study to analyze the cell-type specificity and tissue distribution of the α1, α2 and α5 chains of type IV collagen in human tissues. The results confirmed previous work on α1(IV) and demonstrated that the α2(IV) is codistributed with α1(IV) in all BMs whereas the α5(IV) chain has a more restricted pattern which clearly differs from that reported for α3(IV) and α4(IV). The α5(IV) chain is confined in the kidney only to the GBM, while in all other tissues studied it was localized rather to the stromal type of extracellular matrix.

Materials and Methods

Preparation of Antibodies
Synthetic peptide antigens were made in an Applied Biosystems 430A peptide synthesizer and purified by HPLC. For the preparation of antibodies, the peptides were coupled to ovalbumin or hemocyanin either with the glutaraldehyde [26] or MBS methods. For the production of α1(IV) chain-specific antibodies, a peptide Cys-Gln-Val-Gln-Glu-Lys-Gly-

Asp-Phe-Ala-Thr (α1(IV)-7S peptide) for residues 247–256 from interruption III in the 7S region [27] and an extra cysteine was coupled to hemocyanin and a peptide Cys-Ala-Pro-Ile-Thr-Gly-Glu-Asn-Ile-Arg (α1(IV)-NC peptide) for residues 1534–1542 from the non-collagenous NC1 domain [28] was coupled to ovalbumin by the MBS method. For the α2(IV) chain[29], peptide Cys-Asp-Thr-Asp-Val-Lys-Arg-Ala-Val-Gly-Gly-Asp-Arg-Gln-Glu (α2(IV)-loop peptide) for residues 662–676 from the loop-structured interruption XIII was coupled to hemocyanin by the MBS method, and peptide Gln-Ser-Phe-Gln-Gly-Ser-Pro-Ser-Ala-Asp (α2(IV)-NC peptide) for residues 1682–1691 from the NC domain was coupled to ovalbumin with the glutaraldehyde method. For the α5(IV) chain, peptide Ser-Asp-Met-Phe-Ser-Lys-Pro-Gln-Ser-Glu (α5(IV)-NC peptide) for residues 1655–1664 from the NC domain region was coupled to ovalbumin with the glutaraldehyde method as previously reported [3].

Rabbits were immunized by several subcutaneous injections, with 1 mg peptide-carrier conjugate in 0.5 ml phosphate-buffered saline (PBS) mixed with an equal volume of Freund's complete adjuvant (Difco). Booster injections with the same amount of antigens in Freund's incomplete adjuvant (Difco) were given subcutaneously at 2- to 4-week intervals. The highest titers of antisera were obtained after 7–9 booster injections.

Primary screening for specificity and titer of the antisera was performed by dot-blot assays against the peptide antigens with different carriers and against the corresponding conjugate for different chains. Specificity of the antibodies was verified by Western blot analysis against HT 1080 cell extracts electrophoresed on 6% SDS-PAGE and transferred to a nitrocellulose filter. For demonstration of chain specificity the antisera were preincubated with 100 μg/2 ml of 1:100 diluted antiserum to study blocking of immunostaining.

Tissue Processing
The following tissues were analyzed; brain, eye, ear, thyroid gland, trachea, lung, aorta, heart, liver, spleen, gut, adrenal gland, kidney, muscle, skin, and placenta. The tissues were taken at ten autopsies of children and adults. About 1×1 cm^3 pieces of tissue were frozen in liquid nitrogen, cut at 7 μm, air dried, fixed in cold (-20 °C) acetone 2 min or cold methanol for 5 min or 10% neutral buffered formalin for 20 min, washed twice for 15 min in PBS, and cryopreserved at -20 °C. Eye bulbs were first fixed in 10% neutral buffered formalin, washed twice for 15 min in PBS, frozen in liquid nitrogen, and thereafter cut at 7 μm. Inner ears were fixed in 10% formalin, decalcified in 0.2% EDTA, washed twice for 15 min in PBS, frozen in liquid nitrogen, and cut at 7 μm.

Immunohistochemistry
The tissue samples were stained by the indirect immunofluorescence technique. The fixed cryostat sections wre first incubated in 10% horse serum that was blotted away, then kept overnight at +4 °C in the primary antibody diluted at 1:100, followed by three washes in PBS each for 10 min. The samples were then incubated for 30 min in the tetramethylisothio-cyanate-conjugated secondary antibody against rabbit immunoglobulins (Jackson Laboratories), followed by three washes in PBS, and mounting in glycerol-PBS. A Leitz microscope, equipped with epifluorescence, was used for the fluorescence microscopy. In the controls the primary antibody was omitted or replaced with preimmune serum. For inhibition studies some of the antibodies were preadsorbed overnight with the corresponding peptide conjugate.

If the primary antibody was omitted or preimmune serum was used in its place, no staining was obtained. The antibodies preadsorbed with the corresponding peptide conjugate antigen did not stain the tissues at 1:100 dilution.

Results

Specificity of Antibodies
The α5(IV)-NC peptide antiserum has previously been shown not to react with the α1(IV) or α2(IV) chains [3]. The specificity of the two antisera made here against the α1(IV) chain and the two antisera against the α2(IV) chain was analyzed by Western blotting against HT1080 cell extracts electrophoresed on 5% polyamide. The α1(IV)-NC and α1(IV)-7S antisera both stained a single band of approximately 185 kD (fig. 1, lanes 2 and 6) which corresponds to the size of the α1(IV) chain [30] and the reactivity of both antisera could be blocked by preincubation with the respective antigens (lanes 3 and 7). The staining of the α1(IV)-NC antiserum could not be blocked with an α5(IV) chain antigen (α5(IV)-NC) made from the corresponding region in the α5(IV) chain with 50% sequence identity with the α1(IV)-NC peptide sequence. This indicates that the α1(IV)-NC peptide antiserum is specific for the α1(IV) chain. The α1(IV)-7S antiserum reaction was not affected by a nonspecific peptide containing a 10-residue sequence from the NC domain.

The antisera raised against the α2(IV)-NC peptide with sequence from the NC domain and α2(IV)-coll-loop peptide with sequence from the loop structure in interruption XIII of the collagenous domain both gave a strong reaction with a single band of approximately 170 kD (fig. 1, lanes 10 and 14) which corresponds to the size of α2(IV) [30]. The reaction could be blocked by preincubation of the antisera with the respective antigens (fig. 1, lanes 11 and 15). The staining of the two α2(IV) peptide antisera was not affected by preincubation with a nonspecific peptide α1(IV)-NC with sequence from the NC domain of the α1(IV) chain.

Immunohistochemical Analyses
The tissue localization of α1(IV), α2(IV), and α5(IV) collagen chains was evaluated by the indirect immunofluorescence method in fresh frozen sections of human tissues. The polyclonal antibodies against the α1(IV) and α2(IV) peptides reacted with the tissue sections fixed with either acetone, methanol, or 10% formalin. Fixation with either methanol or acetone preserved the reactivity of the antibodies against the α5(IV) peptide, but fixation with 10% formalin abolished the staining reaction with both antibodies against the α5(IV) peptide. Therefore, eye and ear tissues could not be analyzed by these

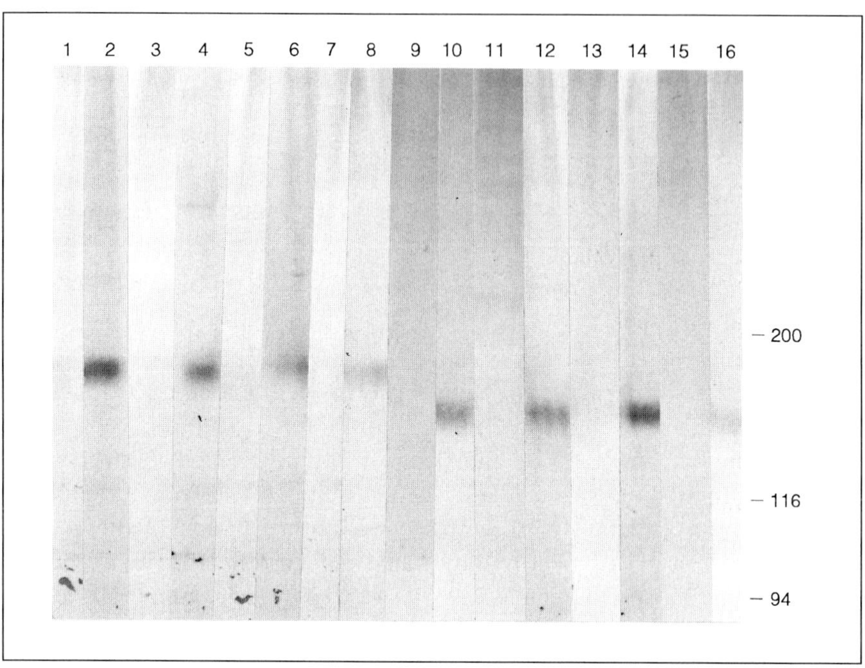

Fig. 1. Immunoblot analyses of antisera raised against human α1(IV) and α2(IV) chain synthetic peptides. Total cellular protein from HT1080 cells was electrophoresed on 5% SDS-PAGE, transferred to nitrocellulose strips and stained with antiserum antisera (see Materials and Methods). Lanes: 1, preimmune serum; 2, α1(IV)-7S antiserum detects a 185-kD band; 3, preincubation of antiserum with the antigen (100 µg/2 ml of 1:100 diluted antiserum) blocks staining; 4, preincubation of antiserum with an α2(IV)-NC peptide conjugate does not block the staining reaction; 5, preimmune serum; 6, α1(IV)-NC antiserum; 7, α1(IV)-NC antiserum preincubated with the antisera; 8, α1(IV)-NC antiserum preincubated with the α5(IV)-NC peptide from the corresponding region of the α5(IV) chain; 9, preimmune serum; 10, α2(IV)-coll-loop antiserum; 11, α2(IV)-coll-loop antiserum preincubated with antisera; 12, α2(IV)-coll-loop antiserum preincubated with α1(IV)-NC peptide; 13, preimmune serum; 14, α2(IV)-NC antiserum; 15, α2(IV)-NC antiserum preincubated with antisera; 16, α2(IV)-NC antiserum preincubated with α1(IV)-NC peptide.

antibodies. All further analyses described in this study were done in acetone-fixed frozen sections.

The immunostaining results demonstrated that the α1(IV) and α2(IV) chains are distributed in all BMs of the human tissues studied (table 1). Two different antisera made for each chain gave identical results. Staining with the antiserum against the α5(IV) chain gave, in general, a totally different result

Table 1. Expression of α1(IV), α2(IV), and α5(IV) chains of type IV collagen in human tissues.

	α1	α2	α5
Brain Vascular BMs	+ +	+ +	−
Thyroid gland Follicular BM	+ + +	+ + +	−
Lung Alveolar BM	+ + +	+ + +	−
Perialveolar stroma	−	−	+ +
Heart Muscle	−	−	−
Liver Sinusoidal BM	+	+ +	−
Perihepatocytic area	−	−	(+)
Portal BM	+ + +	+ + +	−
Pancreas Ductal BM	+ + +	+ + +	−
Islets	+ +	+ +	−
Stroma	−	−	+
Intestine Epithelial BM	+ + +	+ + +	−
Kidney Tubular BM	+ + +	+ + +	−
Glomerular BM	+ + +	+ + +	+ + +
Testis Seminiferous tubules	+ + +	+ + +	−
Ovary Follicles	+ + +	+ + +	−
Spleen Red bulb	+ +	+ +	+ +
White bulb	+	+	−
Muscle Myocytes	+ + +	+ + +	−
Skin Epidermal BM	+ + +	+ + +	−
Dermal stroma	−	−	(+)

The BM of the vascular endothelium is constantly labeled by the antibodies against α1(IV) and α2(IV), but not by the antibodies against α5(IV). Eye and ear are not listed as only formalin-fixed tissues were available and the antibodies against α5(IV) showed poor reactivity after formalin fixation. Faint, questionable staining is shown within parentheses.

than that observed with the α1(IV) and α2(IV) chain antisera. Although localized to several organs, the α5(IV) chain was usually not visualized in the BMs themselves, but rather in the stromal-type extracellular matrix, in particular in the dense connective tissue of the organs listed in table 1. However, in contrast to the stromal localization of the α5(IV) chain in most tissues, the BM of kidney glomeruli was strongly labeled by the antibodies against the α5(IV) chain, as we have previously reported (fig. 2, ref. 3). No differences were found in the expression of the three collagen chains in adult or neonatal organs.

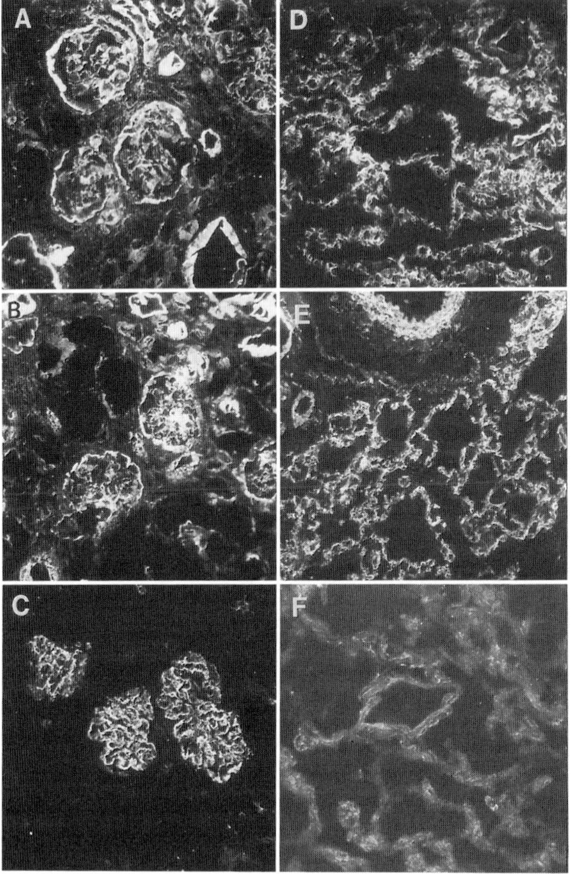

Fig. 2. Immunostaining of human kidney (*A–C*) and lung (*D–F*) with α1(IV) (*A, D*), α2(IV) (*B, E*) and α5(IV) (*C, F*) chain-specific antibodies. The α1(IV) and α2(IV) antibodies show equally strong, even staining of all epithelial and endothelial BMs in both tissues. In contrast, the α5(IV) antibody gives stromal staining pattern lung samples, in particular in the perialveolar stroma. In kidney cortex only the GBM is stained.

The pattern of expression of the three different collagen chains in some human organs is illustrated in figures 2 and 3. In kidney cortex the α1(IV) and α2(IV) chains were visualized in all BMs, i.e. those of glomeruli, Bowman's capsule, tubuli and blood vessels (fig. 2A,B). This was contrasted by the localization of the α5(IV) chain solely to the GBM (fig. 2C). In the lung the α1(IV) and α2(IV) chains were similarly located to epithelial (alveolar and bronchial) and vascular (endothelial) BMs while the α5(IV) chain showed

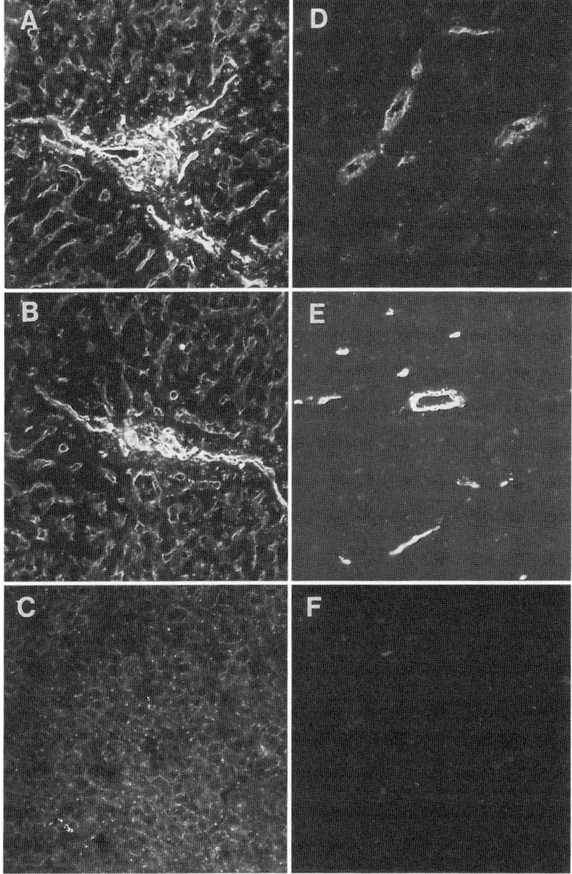

Fig. 3. Immunostaining of human liver (*A–C*) and brain (*D–F*) with α(IV) (*A, D*), α2(IV) (*B, E*) and α5(IV) (*C, F*) chain-specific antibodies. The α1(IV) and α2(IV) antibodies label strongly the vascular, portal and sinusoidal BMs in liver, but the antibody against α5(IV) chain labels very faintly the area around hepatocytes, but not the BMs. In brain, only α1(IV) and α2(IV) chains are expressed.

moderate staining in the stromal compartment of the alveolar wall (fig. 2D–F). In liver both α1(IV) and α2(IV) chains were expressed in sinusoidal and vascular BMs (fig. 3A,B), but α5(IV) chain was faintly expressed around the hepatocytes, but not in the sinusoidal or vascular BMs (fig. 3C). In the brain the α1(IV) and α2(IV) chains were colocalized in small vascular BMs, and α5(IV) was not detected (fig. 3D–F). In spleen, α1(IV) and α2(IV) chains were expressed in the vascular walls both in red and white bulb, but α5(IV) chain was seen in a fibrillar pattern only in red bulb. In skin the α1(IV) and α2(IV) chains were localized to the epithelial and vascular BMs, whereas the α5(IV) antibody showed no signal in these BMs, but a faint, questionable expression was seen in the stromal cells.

Discussion

Recent data have revealed the existence of at least six genetically distinct type IV collagen α chains, demonstrating that this collagen exists in several isoforms. We have now analyzed three different α chains of IV collagen in human tissues, by generating and using chain-specific antibodies made against synthetic peptides. It was demonstrated by immunostaining with the α1(IV) and α2(IV) chain-specific antibodies that these chains are ubiquitous BM components with identical localizations. In contrast, the only BM present in the α5(IV) chain was expressed in abundance in the GBM. Faint α5(IV) chain expression was also seen in the stromal extracellular matrix, in particular, of lung and spleen.

The results of this study showed that the α5(IV) chain has a distribution quite different from that reported for the α3(IV) and α4(IV) chains [11, 12]. Thus, we confirmed previous findings [3] that the α5(IV) chain is present in the kidney only in the GBM, while the α3(IV) and α4(IV) chains are located also in the vascular BMs, Bowman's capsule and TBM [11, 12]. A more striking difference was that, as opposed to the α3(IV) and α4(IV) chains present in several extrarenal BMs, the α5(IV) chain, although present in a variety of extrarenal tissues, is present in only small amounts and usually in the stromal matrix, not in the BMs themselves. Type IV collagen is certainly the major structural component of BMs, and almost exclusively found in these structures, and synthesized by epithelial and other BM-associated cells, but it can be expressed in some stromal-type cells as well. For example, in a recent study employing interspecies chimeras and species-specific antibodies as well as in situ hybridization, it was shown that type IV collagen of BMs around tumor xenografts is synthesized both by epithelial and stromal cells [30]. By in situ hybridization, expression of α1(IV) has similarly been found in stromal and epithelial cells of placenta [31] and, furthermore, its expression is significant in stromal cells of ovarial carcinoma not in the epithelial or cancer cells [32]. It is possible that the normally strong accumulation of α1(IV) and α2(IV) in the BMs renders it difficult to visualize traces of these polypeptides in the stroma. Kleppel et al. [33] have shown that an Alport alloantibody putatively directed against the α5(IV) chain stains normal skin epithelial BM which contradicts our present results. It is possible that the epitope detected by our α5(IV) chain antibody is masked in the skin BM, or alternatively the Alport antigen alloantibody is directed against another protein.

To date, the chain composition of type IV collagen molecules containing the α3(IV), α4(IV) and α5(IV) is not known. However, the practically identical restricted distribution of α3(IV) and α4(IV) indicated that they are present in the same heterotrimer molecules. Analyses on collagenase-solubilized NC domains from GBM support this hypothesis [13, 21]. In contrast, the results of this study indicate that the α5(IV) chain is usually not present in the same molecule as α1(IV), α2(IV), α3(IV) and α4(IV) except in the GBM where all chains are present. Therefore, it seems likely that α5(IV) is usually present in homotrimers or possibly in heterotrimers associated with the recently identified α6(IV) chain whose gene colocated with the COL4A5 gene on the X-chromosome. The GBM is formed through fusion of two BMs synthesized by opposite cell types. It is possible that α5(IV) has a function in the integration of the fusing epithelial and endothelial BMs. This proposal is supported by the late expression of the α5(IV) chain during the cap stage of glomerulogenesis, when the BMs first fuse [K. Sainio and K. Tryggvason, unpubl. results], and also by the ultrastructural findings in Alport syndrome where GBM is split with lamellations and patchy thinning [34]. A large number of families with Alport syndrome has been shown to have aberrations in the α5(IV) gene [16–18]. In other organs, such as lung and spleen, the α5(IV) chain is found only as a minor component in the stromal extracellular matrix. This, in turn, may explain why these organs are not affected in Alport syndrome as opposed to Goodpasture syndrome characterized by both hematuria and lung hemorrhage and α3(IV) autoantibodies. The expression of the α5(IV) chain could not be analyzed in the ear or eye, because the antibodies did not react in the formalin-fixed tissue after decalcification. These tissues, which also are affected in Alport syndrome, remain to be explored by other methods. The question still remains why the α3(IV) and α4(IV) antigens which are the products of genes on chromosome 2 are absent in some patients with X-linked Alport syndrome? This could be explained by the α3(IV), α4(IV) and α5(IV) being located in one heterotrimeric molecule so that the absence of the α5(IV) chain, due to a defect in its gene, would lead to intracellular degradation of the α3(IV) and α4(IV) chains. However, the present results showing different distribution of the α5(IV) chain on the one hand and α3(IV) and α4(IV) on the other do not directly support this hypothesis. Future analysis of the distribution of the recently identified α6(IV) chain [9] may shed more light on this controversial subject.

References

1 Hudson BG, Reeders ST, Tryggvason K: Type IV collagen: Structure, gene organization, and role in human diseases: Molecular basis of Goodpasture and Alport syndromes and diffuse leiomyomatosis. Minireview. J Biol Chem 1993;268:26033–26036.

2 Butkowski RJ, Langeveld JPM, Wieslander J, Hamilton J, Hudson BG: Localization of the Goodpasture epitope to a novel chain of basement membrane collagen. J Biol Chem 1987;262:7874–7877.

3 Hostikka SL, Eddy RL, Byers MG, Höyhtyä M, Shows TB, Tryggvason K: Identification of a distinct type IV collagen α chain with restricted kidney distribution and assignment of its gene to the locus of X chromosome-linked Alport syndrome. Proc Natl Acad Sci USA 1990;87:1606–1610.

4 Gunwar S, Saus J, Noelken ME, Hudson BG: Glomerular basement membrane identification of a fourth chain, α4, of type IV collagen. J Biol Chem 1990;265:5466–5469.

5 Morrison KE, Mariyama M, Yang-Feng TL, Reeders ST: Sequence and localization of a partial cDNA encoding the human α3 chain of type IV collagen. Am J Hum Genet 1991;49:545–554.

6 Kamagata Y, Mattei MG, Ninomiya Y: Isolation and sequencing of cDNAs and genomic DNAs of basement membrane collagen type IV and assignment of the gene to the distal long arm of human chromosome 2. J Biol Chem 1992;267:23753–23758.

7 Mariyama M, Kalluri R, Hudson BG, Reeders ST: The α4(IV) chain of basement membrane collagen. Isolation of cDNAs encoding bovine α4(IV) and comparison with other type IV collagens. J Biol Chem 1992;267:1253–1258.

8 Butkowski RJ, Wieslander J, Kleppel M, Michael AF, Fish AJ: Basement membrane collagen in the kidney: Regional localization of novel chains related to collagen IV. Kidney Int 1989;35:1195–1202.

9 Zhou J, Mochizuki T, Smeets H, Antignac C, Laurila P, de Paepe A, Tryggvason K, Reeders ST: Deletion of the paired α5(IV) and α6(IV) collagen genes in inherited smooth muscle tumors. Science 1993;261:1167–1169.

10 Kleppel MM, Kashtan C, Santi PA, Wieslander J, Michael AF: Distribution of familial nephritis antigen in normal tissue and renal basement membranes of patients with homozygous and heterozygous Alport familial nephritis. Relationship of familial nephritis and Goodpasture antigens to novel collagen chains and type IV collagen. Lab Invest 1989;61:278–289.

11 Kleppel MM, Santi PA, Cameron JD, Wieslander J, Michael AF: Human tissue distribution of novel basement membrane collagen. Am J Pathol 1989;134:813–825.

12 Kleppel MM, Michael AF: Expression of novel basement membrane components in the developing human kidney and eye. Am J Anat 1990;187:165–174.

13 Johansson C, Butkowski R, Wieslander J: The structural organization of type IV collagen. Identification of three NC1 populations in the glomerular basement membrane. J Biol Chem 1992;267:24533–24537.

14 Sanes JR, Engvall E, Butkowski R, Hunter DD: Molecular heterogeneity of basal laminae: Isoforms of laminin and collagen IV at the neuromuscular junction and elsewhere. J Cell Biol 1990;111:1685–1699.

15 Shen G-Q, Butkowski R, Cheng T, Wieslander J, Katz A, Cass J, Fish AJ: Comparison of noncollagenous type IV collagen subunits in human glomerular basement membrane, alveolar basement membrane, and placenta. Connect Tissue Res 1990;24:289–301.

16 Barker DF, Hostikka SL, Zhou J, Chow LT, Oliphant AR, Gerken SC, Gregory MC, Skolnick MH, Atkin CL, Tryggvason K: Identification of mutations in the COL4A5 collagen gene in Alport syndrome. Science 1990;248:1224–1227.

17 Zhou J, Barker D, Hostikka SL, Gregory M, Atkin C, Tryggvason K: Single base mutation in α5(IV) collagen chain gene converting a conserved cysteine to serine in Alport syndrome. Genomics 1991;9:10–18.

18 Tryggvason K, Zhou J, Hostikka SL, Shows T: Molelcular genetics of Alport syndrome. Kidney Int 1993;43:38–44.

19 Wieslander J, Barr JF, Butkowski R, Edwards SJ, Bygren P, Heinegård D, Hudson BG: Goodpasture antigen of the glomerular basement membrane: Localization to noncollagenous regions of type IV collagen. Proc Natl Acad Sci USA 1984;81:3838–3842.

20 Gunwar S, Noelken ME, Hudson BG: Properties of the collagenous domain of the α3(IV) chain, the Goodpasture antigen, of lens basement membrane collagen. Selective cleavage of α(IV) chains with retention of their triple helical structure and noncollagenous domain. J Biol Chem 1991;266: 14088–14094.

21 Hudson BG, Wieslander J, Wisdon BJ Jr, Noelken ME: Biology of disease. Goodpasture syndrome: Molecular architecture and function of basement membrane antigen. Lab Invest 1989;61:256–269.

22 Kalluri R, Gunwar S, Reeders ST, Morrison KC, Mariyama M, Ebner KE, Noelken ME, Hudson BG: Goodpasture syndrome. Localization of the epitope for the autoantibodies to the carboxyl-terminal region of the α3(IV) chain of basement membrane collagen. J Biol Chem 1991;266: 24018–24024.

23 Saus J, Wieslander J, Langeveld J, Quinones S, Hudson BG: Identification of the Goodpasture antigen as the α3(IV) chain of collagen IV. J Biol Chem 1988;263:13374–13380.

24 Kleppel MM, Kashtan CE, Butkowski RJ, Fish AJ, Michael AF: Alport familial nephritis. Absence of 28 kilodalton non-collagenous monomers of type IV collagen in glomerular basement membrane. J Clin Invest 1987;80:263–266.

25 Mariyama M, Zheng K, Yang-Feng TL, Reeders ST: Colocalization of the genes for the α3(IV) and α4(IV) chains of type IV collagen to chromosome 2 bands q35–q37. Genomics 1992;13:809–813.

26 Kagan A, Glick M, in Jaffe BB, Behrman HD (eds): Methods of Hormone Radioimmunoassay. New York, Academic Press, 1979, pp 328–329.

27 Soininen R, Haka-Risku T, Prockop DJ, Tryggvason K: Complete primary structure of the α1 chain of human basement membrane (type IV) collagen. FEBS Lett 1987;225:188–194.

28 Pihlajaniemi T, Tryggvason K, Myers J, Kurkinen M, Lebo R, Cheung M, Prockop DJ, Boyd CD: cDNA clones for the pro-α1(IV) chain of human type IV procollagen reveal an unusual homology of amino acid sequences in two halves of the carboxyl-terminal domain. J Biol Chem 1985;260: 7681–7687.

29 Hostikka SL, Tryggvason K: The complete primary structure of the α2 chain of human type IV collagen and comparison with the α1(IV) chain. J Biol Chem 1988;263:19488–19493.

30 Tryggvason K, Gehron Robey P, Martin GR: Biosynthesis of type IV procollagens. Biochemistry 1980;19:1284–1289.

31 Autio-Harmainen H, Sandberg M, Pihlajaniemi T, Vuorio E: Synthesis of laminin and type IV collagen by trophoblastic cells and fibroblastic stromal cells in the early human placenta. Lab Invest 1991;64:483–491.

32 Autio-Harmainen H, Hurskainen T, Niskasaari K, Höyhtyä M, Tryggvason K: Simultaneous expression of 70 kilodalton type IV collagenase and type IV collagen α1(IV) chain genes by cells of early human placenta and gestational endometrium. Lab Invest 1992;67:191–200.

33 Kleppel MM, Fan WW, Cheong HI, Michael AF: Evidence for separate networks of classical and novel basement membrane collagen. J Biol Chem 1992;267:4137–4142.

34 Atkin CL, Gregory MC, Border WA: Alport syndrome; in Schrier RW, Gottschalk CW (eds): Diseases of Kidney. Boston, Little, Brown, 1988, pp 617–641.

Karl Tryggvason, MD, PhD, Department of Medical Biochemistry and Biophysics, Karolinska Institute, S-171 77 Stockholm (Sweden)

Tryggvason K (ed): Molecular Pathology and Genetics of Alport Syndrome.
Contrib Nephrol. Basel, Karger, 1996, vol 117, pp 142–153

Immunohistologic Findings in Alport Syndrome

Clifford E. Kashtan[a], *Mary M. Kleppel*[a], *Marie-Claire Gubler*[b]

[a] Department of Pediatrics, University of Minnesota Medical School, Minneapolis,
Minn., USA, and
[b] Hôpital Necker – Enfants Malades, INSERM U423, Paris, France

Immunologic techniques (immunohistology and immunochemistry) have
been of key importance in the delineation of the Alport defect. From the
seminal observations of Olson and McCoy and their colleagues in the late
1970s, to current studies of basement membrane composition in Alport
patients (see below), data obtained using these techniques have guided and
inspired investigators in this field. In this chapter we will review results of
immunohistologic and immunochemical studies in Alport syndrome, and
discuss how these sorts of studies can continue to provide insights into the
pathogenesis of this disease.

The modern era of Alport research can be reasonably said to have begun
with electron microscopic studies of Alport glomeruli reported in the early
1970s by several groups of investigators [1–3]. These studies identified unique
alterations of glomerular basement membrane (GBM) structure in patients
with Alport syndrome, and focused attention on GBM as the primary site of
the Alport lesion. Olson et al. [4] and McCoy et al. [5] provided the first solid
clue to the biochemical nature of this lesion by demonstrating that Alport
GBM failed to bind human anti-GBM (Goodpasture) autoantibodies, suggesting the absence of a normal basement membrane constituent from Alport
GBM. In 1984, Wieslander et al. [6] localized the Goodpasture epitope to
type IV collagen, establishing this protein as the target of subsequent Alport
research. Subsequently the Goodpasture epitope was identified as the carboxyterminal noncollagenous (NCI) domain of a new type IV collagen chain, i.e.
$\alpha3(IV)$ [7]. Studies by Savage et al. [8] and by Gubler et al. [9] employing a
monoclonal antibody directed against the Goodpasture antigen confirmed the

absence of this epitope from the GBM of most Alport males. McCoy et al. [10] had demonstrated in 1982 that Alport patients could generate antibodies against the GBM of a renal allograft. Kashtan et al. [11] showed in 1986 that such antibodies were directed against epitopes of type IV collagen, supporting an Alport-type IV collagen connection. This connection was finally established by the cloning of COL4A5, the gene encoding the $\alpha 5$ chain of type IV collagen, by Hostikka et al. [12], and the demonstration by Barker et al. [13] of mutations in this gene in kindreds with Alport syndrome.

Collagenous Constituents of Basement Membranes

Basement membranes contain at least six genetically distinct chains of type IV collagen, as discussed in chapters 3 and 4. While the $\alpha 1$ and $\alpha 2$ chains of type IV collagen are found ubiquitously in basement membranes, the $\alpha 3$, $\alpha 4$, and $\alpha 5$ chains of type IV collagen exhibit a restricted distribution (table 1) [14–16]. Interestingly, those basement membranes known or thought to be abnormal in Alport syndrome (GBM, lens capsule, internal limiting membrane of the retina, Descemet's membrane, and basement membranes of the cochlea) normally express the $\alpha 3$, $\alpha 4$ and $\alpha 5$ chains of type IV collagen. In the glomerulus the $\alpha 3$, $\alpha 4$ and $\alpha 5$(IV) chains are absent from the mesangium, in contrast to the $\alpha 1$ and $\alpha 2$ chains (fig. 1).

GBM also contains two other collagens, types V and VI collagen [17, 18]. These collagens are normally found just beneath the glomerular endothelium [19]. They are also found in the glomerular mesangium and in the renal interstitium [17, 18].

Immunohistology of Alport Basement Membranes

As noted above, Olson et al. [4] and McCoy et al. [5] observed that Alport GBM did not bind anti-GBM autoantibodies in sera from patients with Goodpasture syndrome or isolated anti-GBM nephritis (table 2). Reports by Jenis et al. [20] and Noel et al. [21] indicated that this was not a universal phenomenon among Alport patients, however. The paper by Jenis suggested that expression of Goodpasture epitopes was more likely in females with Alport syndrome, and this was supported by the study of Jeraj et al. [22]. About this time, O'Neill et al. [23] at the University of Utah had published a strong argument contending that Alport syndrome was an X-linked dominant disorder. X-linked dominant inheritance provided an explanation for the observations of Jenis and Jeraj, since female heterozygotes could be expected to

Table 1. Type IV collagen chains

Chain	Gene	Chromo-some	Distribution		
			kidney	eye	cochlea
α1(IV)	COL4A1	13	mesangium GBM (subendothelial) Bowman's capsule all TBM and vessels	ubiquitous	ubiquitous
α2(IV)	COL4A2	13	mesangium GBM (subendothelial) Bowman's capsule all TBM and vessels	ubiquitous	ubiquitous
α3(IV)	COL4A3	2	GBM (lamina densa) Bowman's capsule distal TBM	lens capsule Descemet's membrane Bruch's membrane internal limiting membrane	inner/outer sulci spiral limbus spiral prominence basilar membrane
α4(IV)	COL4A4	2	GBM (lamina densa) Bowman's capsule distal TBM	lens capsule Descemet's membrane Bruch's membrane internal limiting membrane	inner/outer sulci spiral limbus spiral prominence basilar membrane
α5(IV)	COL4A5	X	GBM (lamina densa) Bowman's capsule distal TBM	lens capsule Descemet's membrane Bruch's membrane internal limiting membrane	inner/outer sulci spiral limbus spiral prominence basilar membrane
α6(IV)[a]	COL4A6	X	?	?	?

[a] The distribution of α6(IV) is presently unknown, although studies using antibodies to this protein are in progress. Messenger RNA for α6(IV) has been found in placenta, adult kidney, skeletal muscle and lung.

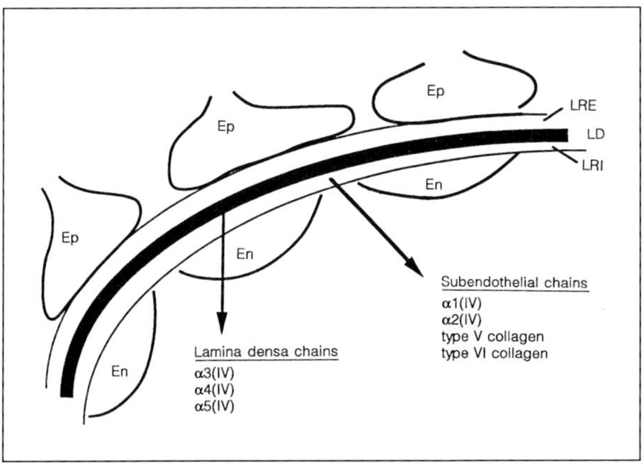

Fig. 1. Collagen chain distribution in normal human GBM.

Table 2. Immunohistological studies of the GBM in Alport syndrome

| Author | Year | GBM immunolabelling | | | | | Antisera and antibodies |
| | | − | | ± | + | | |
		M	F	F	M	F	
McCoy	1976, 1982	7				2	GP/FNS
Olson	1980	5	2				GP
Jenis	1981	2	2			5	GP
Noël	1981				4	2	GP
Jerah	1983	7			8	6	GP
Savage	1986	9	1			1	MCA
Kashtan	1986	5					FNS
Noël	1988		20			22	MCA
Kleppel	1989	8		1			GP/FNS
Thorner	1990	1		1			GP

Studies have been performed using sera from patients affected with Goodpasture syndrome (GP) or from Alport syndrome patients who developed anti-GBM glomerulonephritis after transplantation (FNS). MCA is a monoclonal antibody against the GP epitope.

Table 3. Distribution of the α3 chain of collagen IV in the GBM of 78 patients [from 24]

Disease	Glomerular distribution of the α3(IV) chain		
	+	discontinuous	−
Alport's syndrome	17M		32M
(70 patients)	6F	13F	2F
Progressive hereditary	5M		
nephritis without deafness (8 patients)	3F		

display a mosaic pattern of Goodpasture epitope expression, while basement membranes of male hemizygotes would uniformly fail to express such epitopes, assuming that abnormal expression of Goodpasture epitopes was in some way fundamentally related to the pathogenesis of Alport syndrome. Kashtan et al. [11] lent further support to this concept, showing that heterozygotes typically exhibited mosaic expression of a type IV collagen-related epitope in their epidermal basement membranes (EBMs), while EBMs of their affected male relatives showed no expression of the epitope at all.

The dissection of type IV collagen into distinct chains and the development of monoclonal antibodies specific for each chain have allowed investigators to refine their descriptions of collagen chain expression in Alport basement membranes. It is now clear that the α1 and α2 chains of type IV collagen are expressed in Alport basement membranes; in fact, these chains appear to be overexpressed in Alport GBM (see below). On the other hand, GBM of Alport hemizygotes (affected males), frequently exhibit no expression of the α3, α4 and α5 chains of collagen IV. In an immunohistochemical study at the Hôpital Necker of 78 patients with progressive hereditary nephritis, 70 with deafness and 8 with normal hearing, abnormal distribution of the α3 chain of type IV collagen was observed in 67% of those with progressive hereditary nephritis and deafness. The α3 epitope was completely absent from the GBM of male patients whereas discontinuous distribution of the epitope was observed in heterozygous females. In contrast, a third of the patients with progressive hereditary nephritis and deafness and all patients with progressive hereditary nephritis without deafness were found to have normal distribution of the α3(IV) chain (table 3, fig. 2) [24]. To this point, it appears that there is concordance of expression, or lack of expression, of α3(IV), α4(IV) and α5(IV) in males with the X-linked dominant form of Alport syndrome [25, 26]. Individuals with autosomal recessive Alport syndrome exhibit a different pattern of type IV collagen expression: while α3(IV), α4(IV) and α5(IV) are all absent from GBM,

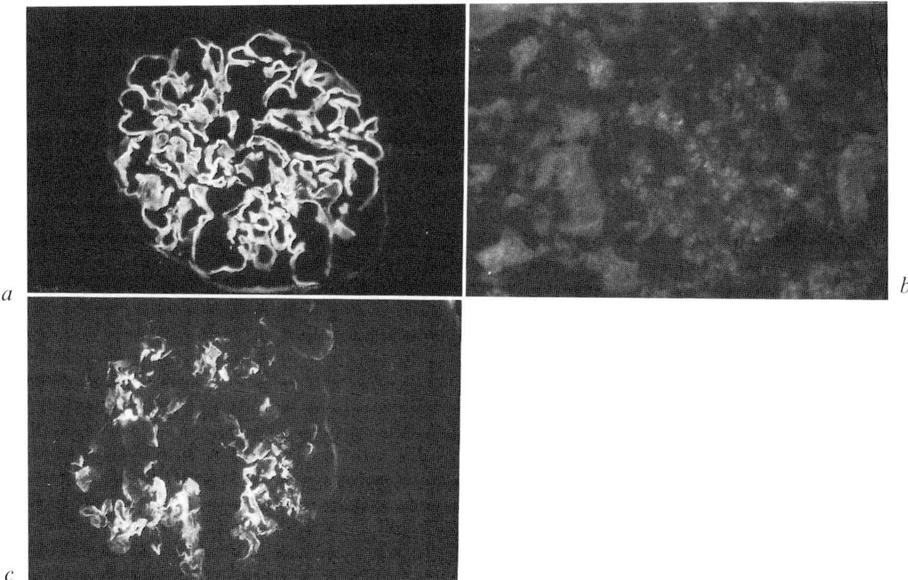

Fig. 2. Kidney biopsy specimens from a normal control (*a*), a male (*b*) and a female (*c*) patient with X-linked Alport syndrome were stained with a monoclonal antibody directed against the NC1 domain of the α3 chain of type IV collagen.

and α3(IV) and α4(IV) are absent from Bowman's capsule and distal TBM, α5(IV) is expressed in Bowman's capsule, distal TBM and EBM [27] (fig. 3).

Other than EBM, expression of type IV collagen chains in extrarenal basement membranes of Alport patients has received little attention, primarily because of the scarcity of appropriate tissues for study. We have recently examined lens capsule from 2 unrelated Alport males with anterior lenticonus; lens capsule from 1 of these patients did not express α3(IV), α4(IV) or α5(IV), while lens capsule from the other expressed each of these chains [28]. To date, it has not been possible to obtain cochlear tissues for study.

As noted in table 1, the α1 and α2 chains of type IV collagen are normally confined to the subendothelial region of GBM. Alport kidneys biopsied early in the course of the nephropathy and studied by indirect immunofluorescence exhibit expression of these chains throughout the apparent width of the GBM [19]. Similarly, types V and VI collagen, normally found in a subendothelial location, also appear to extend through the width of Alport GBM [19]. Kidneys from patients with diabetic nephropathy, membranoproliferative glomerulo-nephritis and focal segmental glomerulosclerosis did not exhibit these changes in α1(IV), α2(IV), type V collagen and type VI collagen expression.

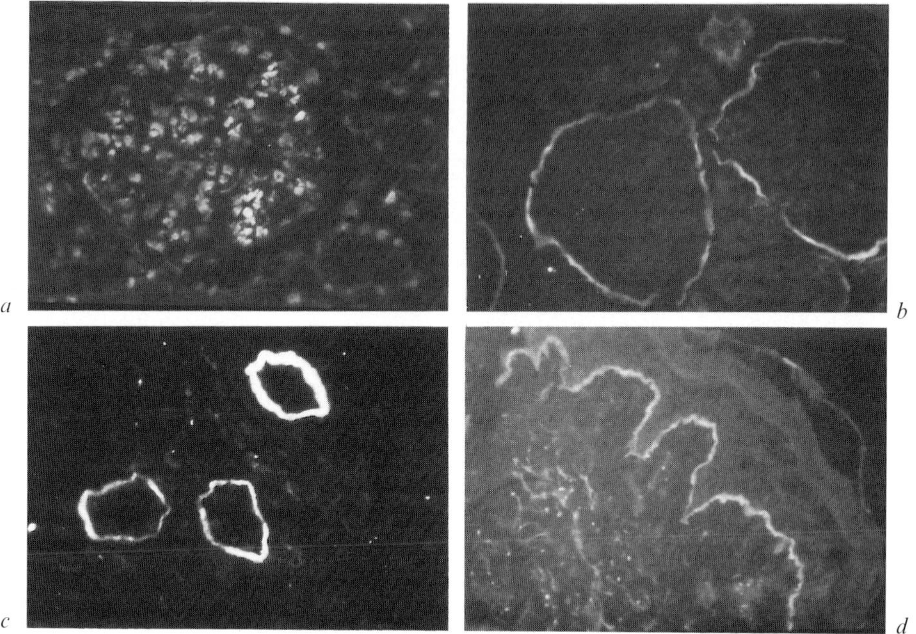

Fig. 3. Kidney (*a–c*) and skin (*d*) biopsy specimens from a male patient affected with autosomal recessive Alport syndrome: α3(IV) distribution (*a*); α5(IV) distribution (*b–d*).

Immunochemistry of Alport Basement Membranes

Immunohistologic descriptions of antigenic abnormalities in Alport glomeruli led to studies of the composition of isolated Alport basement membranes (GBM and tubular basement membrane (TBM)). Kleppel et al. [29] found that Alport GBM and TBM lacked noncollagenous peptides that were targets for anti-GBM autoantibodies. These peptides were subsequently identified as the NC1 domains of the α3 and α4 chains of type IV collagen [7, 30]. Alport GBM and TBM are also lacking in a 26-kD noncollagenous peptide that is reactive with alloantibodies to GBM from Alport patients with posttransplant anti-GBM nephritis [11, 31]. This peptide is specifically recognized by a monoclonal antibody against the NC1 domain of the α5 chain of type IV collagen [32, 33]. Thus, the α3, α4 and α5 chains of type IV collagen are missing from Alport GBM and TBM in most males with X-linked Alport syndrome.

Posttransplant Anti-GBM Nephritis

Anti-GBM nephritis involving the renal allograft is a rare, but dramatic, manifestation of Alport syndrome that was first detected using immunohistologic methods by McCoy et al. [10]. Data from several transplant centers indicate an incidence of 3–4% in transplanted Alport patients [34–38]. If the number of transplanted Alport males is used as the denominator, the incidence is somewhat higher, perhaps on the order of 5–7%. It is important that Alport patients who suffer allograft dysfunction undergo evaluation for anti-GBM nephritis, for several reasons. First, anti-GBM nephritis may respond to appropriate therapy. Second, Alport patients who develop allograft anti-GBM nephritis are very likely to have recurrent anti-GBM nephritis in subsequent allografts [39–44]. Finally, allograft anti-GBM nephritis may occur in more than one member of an Alport kindred [29, 44].

Since the first report of anti-GBM nephritis in the allograft of a patient with Alport syndrome in 1982, over 20 additional cases have been reported or have come to our attention through personal communication [10, 11, 40–48]. A profile of the susceptible patient can be put together by examining 22 cases for which appropriate data is available. The patient is usually male (20/22), always deaf (22/22), and likely to have reached end-stage renal disease before the age of 30 (20/22). These phenotypic characteristics, together with at least two instances of familial occurrence of allograft anti-GBM nephritis, suggest that certain mutations in the COL4A5 gene, such as deletions, may predispose patients to the development of allograft anti-GBM nephritis (see below).

In 87% of these cases, the onset of anti-GBM nephritis occurred within the first year following transplantation, and all of the allografts were eventually lost. Anti-GBM nephritis recurred in 7 of 8 patients who underwent retransplantation. Allograft anti-GBM nephritis may recur despite an interval of many years between transplants, and in the absence of detectable circulating anti-GBM antibodies prior to retransplantation.

Anti-GBM antibodies from these patients appear to fall into two classes. First, there are antibodies that stain EBM as well as GBM, and bind to a 26-kD noncollagenous GBM peptide by Western blotting [11, 31, 34]. This peptide is the noncollagenous domain of the $\alpha 5$ chain of collagen IV [32, 33]. The second class of antibodies does not stain EBM, and by Western blotting resembles Goodpasture antibodies with binding to noncollagenous peptides of 24, 26 and 28 kD molecular size [34, 49, 50]. These antibodies appear to be directed primarily against the NC1 domain of the $\alpha 3(IV)$ chain [51].

It appears that 11–12% of the mutations in the COL4A5 gene causing Alport syndrome are complete or partial deletions [52, and chapter 8]. Preliminary data suggest that the incidence of deletions in patients who develop anti-GBM nephritis after renal transplantation may be significantly greater [48, 52–54]. Deletions in COL4A5 would be more likely to eliminate critical epitopes on the α5(IV) chain, preventing establishment of immunologic tolerance. In this sense, Alport syndrome may be analogous to the hemophilias in which deletions in the genes for factor VIII or IX occur with greater frequency in patients who generate antibodies against transfused coagulation factor [55, 56]. Two females with Alport syndrome and posttransplant anti-GBM nephritis have been shown to have autosomal recessive disease, due to mutations in the COL4A3 gene encoding the α3(IV) chain [57, 58].

References

1 Hinglais N, Grünfeld J-P, Bois LE: Characteristic ultrastructural lesion of the glomerular basement membrane in progressive hereditary nephritis (Alport's syndrome). Lab Invest 1972;27:473–487.
2 Spear GS, Slusser RJ: Alport's syndrome: Emphasizing electron microscopic studies of the glomerulus. Am J Pathol 1972;69:213–222.
3 Churg J, Sherman RL: Pathologic characteristics of hereditary nephritis. Arch Pathol 1973;95:374–379.
4 Olson DL, Anand SK, Landing BH, Heuser E, Grushkin CM, Lieberman E: Diagnosis of hereditary nephritis by failure of glomeruli to bind anti-glomerular basement membrane antibodies. J Pediatr 1980;96:697–699.
5 McCoy RC, Johnson HK, Stone WJ, Wilson CB: Variation in glomerular basement membrane antigen in hereditary nephritis (abstract). Lab Invest 1976;34:325–326.
6 Wieslander J, Barr JF, Butkowski R, Edwards SJ, Bygren P, Heinegard D, Hudson BG: Goodpasture antigen of the glomerular basement membrane: Localization to noncollagenous regions of type IV collagen. Proc Natl Acad Sci USA 1984;81:3838–3842.
7 Saus J, Wieslander J, Langeveld JPM, Quinones S, Hudson BG: Identification of the Goodpasture antigen as the α3(IV) chain of collagen IV. J Biol Chem 1988;263:13374–13380.
8 Savage COS, Pusey CD, Kershaw MJ, Cashman SJ, Harrison P, Harley B, Turner DR, Cameron JS, Evans DJ, Lockwood CM: The Goodpasture antigen in Alport's syndrome: Studies with a monoclonal antibody. Kidney Int 1986;30:107–112.
9 Gubler MC, Mounier F, Gros F, Wieslander J, Beziau A, Guicharnaud L: Glomerular distribution of subunits M1 and M2 of the globular domain of basement membrane collagen in Alport's syndrome; in Gubler MC, Sternberg M (eds): Progress in Basement Membrane Research: Renal and Related Aspects in Health and Disease. London, Libbey Eurotext, 1988, pp 177–182.
10 McCoy RC, Johnson HK, Stone WJ, Wilson CB: Absence of nephritogenic GBM antigen(s) in some patients with hereditary nephritis. Kidney Int 1982;21:642–652.
11 Kashtan C, Fish AJ, Kleppel M, Yoshioka K, Michael AF: Nephritogenic antigen determinants in epidermal and renal basement membranes of kindreds with Alport-type familial nephritis. J Clin Invest 1986;78:1035–1044.
12 Hostikka SL, Eddy RL, Byers MG, Hoyhtya M, Shows TB, Tryggvason K: Identification of a distinct type IV collagen α chain with restricted kidney distribution and assignment of its gene to the locus of X-chromosome-linked Alport syndrome. Proc Natl Acad Sci USA 1990;87:1606–1610.
13 Barker DF, Hostikka SL, Zhou J, Chow LT, Oliphant AR, Gerken SC, Gregory MC, Skolniock MH, Atkin CL: Identification of mutations in the COL4A5 collagen gene in Alport syndrome. Science 1990;248:1224–1227.

14   Butkowski RJ, Wieslander J, Kleppel M, Michael AF, Fish AJ: Basement membrane collagen regional localization of novel chains related to collagen IV. Kidney Int 1989;35:1195–1202.

15   Kleppel MM, Santi PA, Cameron JD, Wieslander J, Michael AF: Human tissue distribution of novel basement membrane collagen. Am J Pathol 1989;134:813–825.

16   Yoshioka K, Hino S, Takemura T, Maki S, Wieslander J, Takekoshi Y, Makino H, Kagawa M, Sado Y, Kashtan CE: Type IV collagen α5 chain: Normal distribution and abnormalities in X-linked Alport syndrome revealed by monoclonal antibody. Am J Pathol 1994;144:986–996.

17   Roll FJ, Madri JA, Albert JA, Furthmayr H: Codistribution of collagen types IV and AB2 in basement membranes and mesangium of the kidney. J Cell Biol 1980;85:597–616.

18   Von der Mark H, Aumailley M, Wick G, Fleischmajer R, Timpl R: Immunochemistry, genuine size and tissue localization of collagen VI. Eur J Biochem 1984;142:493–502.

19   Kashtan CE, Kim Y: Distribution of the α1 and α2 chains of collagen IV and of collagens V and VI in Alport syndrome. Kidney Int 1992;42:115–126.

20   Jenis EH, Valeski JE, Calcagno PL: Variability of anti-GBM binding in hereditary nephritis. Clin Nephrol 1981;15:111–114.

21   Noël L-H, Droz D, Foidart JM, Mahieu PR, Grünfeld J-P: Immunological and biochemical studies of glomerular basement membrane in hereditary nephritis. Proc 8th Int Congr Nephrol, 1981, p 82.

22   Jeraj K, Kim Y, Vernier RL, Fish AJ , Michael AF: Absence of Goodpasture's antigen in male patients with familial nephritis. Am J Kidney Dis 1982;2:626–629.

23   O'Neill WM, Atkin CL, Bloomer HA: Hereditary nephritis: A re-examination of its clinical and genetic features. Ann Intern Med 1978;88:176–182.

24   Gubler MC, Antignac C, Deschênes G, Knebelmann B, Hors-Cayla MC, Grünfeld J-P, Broyer M, Habib R: Genetic, clinical and morphologic heterogeneity in Alport's syndrome. Adv Nephrol 1993; 22:15–35.

25   Kleppel MM, Kashtan C, Santi PA, Wieslander J, Michael AF: Distribution of familial nephritis antigen in normal tissue and renal basement membranes of patients with homozygous and heterozygous Alport familial nephritis: Relationship of familial nephritis and Goodpasture antigens to novel collagen chains and type IV collagen. Lab Invest 1989;61:278–289.

26   Nakanishi K, Yoshikowa N, Iilima K, Kitagawa K, Nakamura H, Ito H, Yoshioka K, Kagawa M, Sado Y: Immunohistochemical study of α1–5 chains of type IV collagen in hereditary nephritis. Kidney Int 1994;46:1413–1421.

27   Gubler MC, Knebelmann B, Beziau A, Broyer M, Pirson Y, Haddoum F, Kleppel MM, Antignac C: Autosomal recessive Alport syndrome: Immunohistochemical study of type IV collagen chain distribution. Kidney Int, in press.

28   Cheong HI, Kashtan CE, Kim Y, Kleppel MM, Michael AF: Immunohistologic studies of type IV collagen in anterior lens capsules of patients with Alport syndrome. Lab Invest 1994;70:553–557.

29   Kleppel MM, Kashtan CE, Butkowski RJ, Fish AJ, Michael AF: Alport familial nephritis: Absence of 28 kilodalton non-collagenous monomers of type IV collagen in glomerular basement membrane. J Clin Invest 1987;80:263–266.

30   Gunwar S, Saus J, Noelken ME, Hudson BG: Glomerular basement membrane: Identification of a fourth chain, α4, of type IV collagen. J Biol Chem 1990;265:5466–4569.

31   Kashtan C, Butkowski R, Kleppel M, First M, Michael A: Posttransplant antiglomerular basement membrane nephritis in related Alport males with Alport syndrome. J Lab Clin Med 1990;116: 508–515.

32   Kleppel MM, Fan W, Cheong H, Kashtan CE, Michael AF: Immunochemical studies of the Alport antigen. Kidney Int 1992;41:1629–1637.

33   Ding J, Kashtan CE, Fan WW, Kleppel MM, Sun MJ, Kalluri R, Neilson EG, Michael AF: A monoclonal antibody marker for Alport syndrome identifies the Alport antigen as the α5 chain of type IV collagen. Kidney Int 1994;46:1504–1506.

34   Van de Heuvel LPWJ, Schröder CH, Savage COS, Menzel D, Assman KJM, Monnes LAH, Veerkamp JH: The development of anti-glomerular basement membrane nephritis in two children with Alport's syndrome after renal transplantation: Characterization of the antibody target. Pediatr Nephrol 1989;3:406–413.

35   Noël L-H, Gubler MC, Bobrie G, Savage COS, Lockwood CM, Grünfeld J-P: Inherited defects
     of renal basement membranes. Adv Nephrol 1989;18:77–94.
36   Berardinelli L, Pozzoli E, Raiteri M, Canal R, Tonello G, Tarantino A, Vegeto A: Renal transplanta-
     tion in Alport's syndrome: Personal experience in twelve patients. Contrib Nephrol. Basel, Karger,
     1990, vol 80, pp 131–134.
37   Peten E, Pirson Y, Cosyns J-P, Squifflet J-P, Alexandre GPJ, Noël LH, Grünfeld J-P, van Ypersele
     de Strihou C: Outcome of thirty patients with Alport's syndrome after renal transplantation.
     Transplantation 1991;52:823–826.
38   Gobel J, Olbricht CJ, Offner G, Helmchen U, Repp H, Koch KM, Frei U: Kidney transplantation
     in Alport's syndrome: Long-term outcome and allograft anti-GBM nephritis. Clin Nephrol 1992;
     38:299–304.
39   Kashtan CE, Atkin CL, Gregory MC, Michael AF: Identification of variant Alport phenotypes
     using an Alport-specific antibody probe. Kidney Int 1989;36:669–674.
40   Milliner DS, Pieredes AN, Holley KE: Renal transplantation in Alport's syndrome: Anti-glomerular
     basement membrane glomerulonephritis in the allograft. Mayo Clin Proc 1982;57:35–43.
41   Shah B, First MR, Mendoza NC, Clyne DH, Alexander JW, Weiss MA: Alport's syndrome:
     Risk of glomerulonephritis induced by anti-glomerular-basement-membrane antibody after renal
     transplantation. Nephron 1988;50:34–38.
42   Rassoul Z, Al-Khader AA, Al-Sulaiman M, Dhar JM, Coode P: Recurrent allograft antiglomerular
     basement membrane glomerulonephritis in a patient with Alport's syndrome. Am J Nephrol 1990;
     10:73–76.
43   Goldman M, Depierreux M, De Pauw L, Vereerstraeten P, Kinnaert P, Noël L-H, Grünfeld J-P,
     Toussaint C: Failure of two subsequent renal grafts by anti-GBM glomerulonephritis in Alport's
     syndrome: Case report and review of the literature. Transplant Int 1990;3:82–85.
44   Helderman JH: The case of the two disparate diseases: A medical mystery. Am J Nephrol 1991;
     11:157–163.
45   Teruel JL, Liano F, Manipaso F, Moreno J, Serrano A, Quereda C, Ortuno J: Allograft anti-
     glomerular basement membrane glomerulonephritis in a patient with Alport's syndrome. Nephron
     1987;46:43–44.
46   Fleming SJ, Savage COS, McWilliam LJ, Pickering SJ, Ralston AJ, Johnson RWG, Ackrill P: Anti-
     glomerular basement membrane antibody-mediated nephritis complicating transplantation in a
     patient with Alport's syndrome. Transplantation 1988;46:857–859.
47   Oliver TB, Gouldesbrough DR, Swainson CP: Acute crescentic glomerulonephritis associated with
     antiglomerular basement membrane antibody in Alport's syndrome after second transplantation.
     Nephrol Dial Transplant 1991;6:893–895.
48   Netzer K-O, Renders L, Zhou J, Pullig O, Tryggvason K, Weber M: Deletions of the COL4A5
     gene in patients with Alport syndrome. Kidney Int 1992;42:1336–1334.
49   Savage COS, Noël L-H, Crutcher E, Price SRG, Grünfeld J-P, Lockwood CM: Hereditary nephritis:
     Immunoblotting studies of the glomerular basement membrane. Lab Invest 1989;60:613–618.
50   Hudson BG, Kalluri R, Gunwar S, Weber M, Ballester F, Hudson JK, Noelken M, Sarras M,
     Richardson WR, Saus J, Abrahamson DR, Glick AD, Haralson MA, Helderman JH, Stone WJ,
     Jacobson HR: The pathogenesis of Alport syndrome involves type IV collagen molecules containing
     the $\alpha3(IV)$ chain: Evidence from anti-GBM nephritis after renal transplantation. Kidney Int 1992;
     42:179–187.
51   Kalluri R, Weber M, Netzer KO, Sun MJ, Neilson EG, Hudson BG: COL4A5 deletion and
     production of post-transplant anti-alpha 3(IV) collagen alloantibodies in Alport syndrome. Kidney
     Int 1994;45:721–726.
52   Antignac C, Knebelmann B, Drouot L, Gros F, Deschênes G, Hors-Cayla MC, Zhou J, Tryggvason
     K, Grünfeld J-P, Broyer M, Gubler MC: Deletions in the COL4A5 collagen gene in X-linked Alport
     syndrome. J Clin Invest 1994;93:1195–1207.
53   Smeets HJM, Melenhorst JJ, Lemmink HH, Schroder CH, Melen MR, Zhou J, Hostikka SL,
     Tryggvason K, Ropers H-H, Jansweijer MCE, Monnens LAH, Brunner HG, van Oost BA: Different
     mutations in the COL4A5 collagen gene in two patients with different features of Alport syndrome.
     Kidney Int 1992;42:83–88.

54 Ding J, Zhou J, Tryggvason K, Kashtan CE: COL4A5 deletions in three patients with Alport syndrome and posttransplant antiglomerular basement membrane nephritis. J Am Soc Nephrol 1994;5:161–168.

55 Gianelli F, Choo KH, Rees DJG, Boyd Y, Rissa CR, Brownlee GG: Gene deletions in patients with haemophilia B and anti-factor IX antibodies. Nature 1983;303:181–182.

56 Millar D, Steinbrecher R, Wieland K, Grundy CB, Martinowitz U, Krawczak M, Zoll B, Whitmore D, Stephenson J, Mibashan RS, Kakkar VV, Cooper DN: The molecular genetic analysis of haemophilia A: Characterization of six partial deletions in the factor VIII gene. Hum Genet 1990; 86:219–227.

57 Mochizuki T, Lemmink HH, Mariyama M, Antignac C, Gubler MC, Pirson Y, Verellen-Dumoulin C, Chan B, Schröder CH, Smeets HJ, Reeders ST: Identification of mutations in the $\alpha3(IV)$ and $\alpha4(IV)$ collagen genes in autosomal recessive Alport syndrome. Nat Genet 1994;8:77–82.

58 Ding J, Stitzel J, Berry P, Hawkins E, Kashtan CE: Autosomal recessive Alport syndrome: Mutation in the COL4A3 gene in a woman with Alport syndrome and posttransplant antiglomerular basement membrane nephritis. J Am Soc Nephrol 1995;5:1–4.

Marie-Claire Gubler, Hôpital Necker – Enfants Malades, INSERM U423,
F–75743 Paris Cedex 15 (France)

Tryggvason K (ed): Molecular Pathology and Genetics of Alport Syndrome.
Contrib Nephrol. Basel, Karger, 1996, vol 117, pp 154–171

Mutations in Type IV Collagen Genes and Alport Phenotypes

Karl Tryggvason

Biocenter Oulu and Department of Biochemistry, University of Oulu, Finland

The real clues to the pathogenic mechanisms of the hereditary Alport syndrome began to evolve in the early 1970s with the advent of electron microscopy and initial characterization of the molecular composition of the glomerular basement membrane (GBM). Electron microscopy studies from several laboratories [1–4] demonstrated ultrastructural changes in the GBM with regional thinning and thickening, usually accompanied by splitting and lamellation. This, in turn, indicated that mutation(s) causing the disease were in some gene or genes encoding protein(s) important for the structure or metabolism of the GBM. In 1971, Kefalides [5] demonstrated for the first time that the GBM contains a unique basement membrane collagen which he referred to as type IV collagen. This collagen, which was believed to be a homotrimer of three $\alpha 1(IV)$ chains, was proposed to form the building block for the GBM scaffold. The identification of this unique structural GBM protein inspired Spear [6] to propose that Alport syndrome might be caused by mutations in the type IV collagen gene. As discussed by Pihlajaniemi (see chapter 3), it turned out, however, that the basement membrane type IV collagen is, in fact, a highly complex collagen subfamily containing at least six genetically distinct α chains, instead of one as was initially believed.

Following the identification and cloning of human cDNAs and genes for the most abundant $\alpha 1(IV)$ and $\alpha 2(IV)$ collagen chains in the mid-1980s [7, 8], and assignment of the genes to chromosome 13 [9], it was evident that these genes could not cause the most common X-linked form of Alport syndrome. Therefore, the attention focused on the novel $\alpha 3(IV)$ and $\alpha 4(IV)$ collagen chains, first identified and partially sequenced by Hudson's group in the late 1980s [10, 11]. The $\alpha 3(IV)$ collagen chain was shown to contain the epitope for Goodpasture syndrome autoantibodies [10] and, interestingly, some Alport syndrome patients did not contain this epitope in their GBM [12]. This suggested that the α3(IV) collagen chain gene was indeed involved in the etiology of Alport syndrome. In 1988, three laboratories independently localized the Alport syndrome locus to Xq22-26 [13–15] and it appeared that the only thing remaining was to isolate the COL4A3 gene and assign it to the X-chromosome. However, this did not turn out to be the case. The real clues to the disease emerged in 1990 when Hostikka et al. [16] and Pihlajaniemi et al. [17] reported the isolation of cDNAs for a novel α5(IV) collagen chain, with gene located at Xq22 [16]. Immediately thereafter, Barker et al. [18] and Zhou et al. [19] reported the first two mutations in COL4A5 in two Utah kindreds. To date, more than 60 mutations have been found in the COL4A5 gene in patients with X-linked Alport syndrome. The COL4A3 and COL4A4 genes which were the initial strong candidates for Alport syndrome were later localized to chromosome 2, and they were also shown to be involved in Alport syndrome, as mutations have recently been found in both genes in the rare autosomal form of the disease [20].

Biosynthesis and Assembly of Type IV Collagen

To understand the consequences of many of the mutations in the COL4A5 gene, it is important to know some of the unique features of collagen biosynthesis (fig. 1) and some basic properties of collagenous proteins. As described in chapter 3, the α chains of all collagens contain numerous consecutive Gly-X-Y repeats, where proline and hydroxyproline are frequently located in positions X and Y. The presence of glycine as every third amino acid is essential, as it is the only amino acid small enough to fit into the center of the triple helix. Transcription of the α chain genes occurs as normally for proteins, but during translation in the cisternae of the rough endoplasmic reticululm the nascent α chains undergo several enzymatic posttranslational modifications. These include hydroxylation at the 4 and 3 positions of prolyl residues by prolyl-4- and prolyl-3-hydroxylases, respectively, hydroxylation of lysyl residues by lysylhydroxylase, and glycosylation of most hydroxylysyl residues by two glucosyl transferases to form glucosyl-galactosyl-hydroxylysyl residues [for review on collagen biosynthesis, see e.g. 21]. Hydroxyproline is essential in collagen where it is required for the maintenance of a stable triple helix through the formation of hydrogen bonds between the α chains. In type IV collagen about 60% of the prolyl residues are hydroxylated to 4-hydroxyproline and about 5% to 3-hydroxyproline, and about 90% of the lysyl residues are hydroxylated with about 90% of them containing glucosyl-galactosyl disaccharides. The hydroxylylsyl residues are believed to stabilize

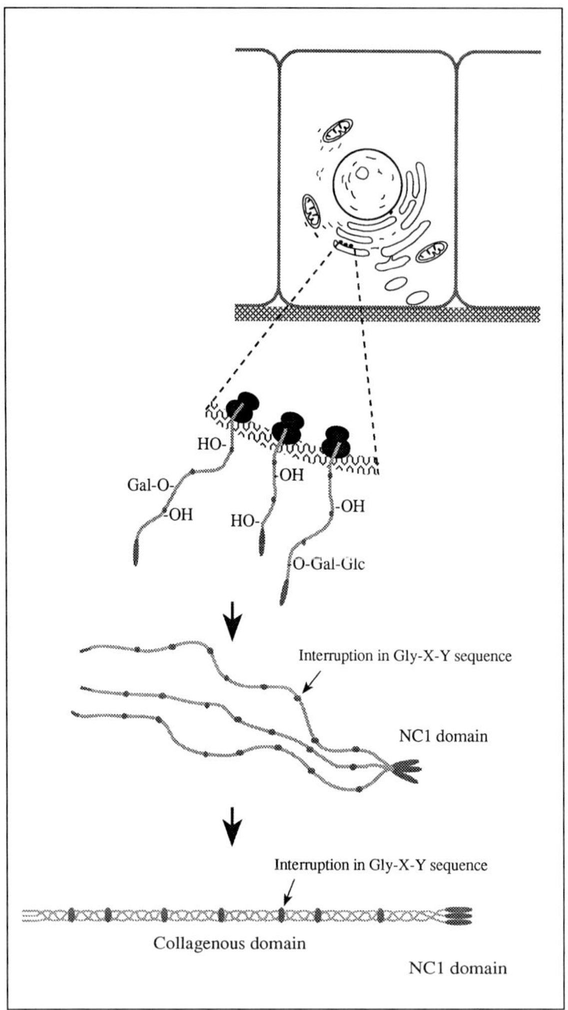

Fig. 1. Schematic summary of the steps of type IV collagen biosynthesis in a cell resting on a basement membrane. Following transcription and processing of the primary transcript, the mRNA is transported to the cytosol to ribosomes where translation occurs. Posttranslational hydroxylations and glycosylations take place on the nascent α chains in the cisternae of the rough endoplasmic reticulum. This is followed by chain association, disulfide bonding and formation of the triple helix of three appropriate α chains. The triple helical molecules are then transported through the Golgi apparatus to the extracellular space where supramolecular assembly occurs as summarized in chapter 3.

cross-links between molecules, but the function of carbohydrates is unknown. Once the triple helix has been formed in the cisternae the posttranslational modifications cease as the enzymes do not react with α chains in a triple helical protein. The triple helical monomers are then secreted out from the cell where they self-assemble into the supramolecular basement membrane network as described in chapter 3.

Mutations in the COL4A5 Gene

To date, numerous reports have described mutations in the COL4A5 gene, the number already exceeding 60, of which 43 had been published when this was written [18, 19, 22–35, 49, 53–58]. The published mutations are listed in tables 1 and 2. The mutations include large and small gene rearrangements such as deletions, insertions, inversions and duplications as well as single base changes. Furthermore, complete loss of the gene has been reported in one case. A striking finding is that the mutations identified so far in COL4A5 are highly dispersed in this huge gene which contains 51 exons and is ~250 kb in size [36, 37]. The same mutation has been found in two apparently unrelated kindreds only in a few cases. This sharply contrasts findings in other diseases such as cystic fibrosis where approximately 70% of the patients have the same mutation [38]. The dispersion of mutations in COL4A5 resembles the situation reported for osteogenesis imperfecta, a brittle bone disease which involves a variety of mutations in type I collagen genes [39].

The mutations present in the COL4A5 gene in Alport syndrome can in most cases well explain the structural and pathophysiological changes in the GBM. The mutations can result in changes that affect: (a) the synthesis of the primary transcript; (b) posttranscriptional modification of the transcript such as splicing; (c) translation of the mRNA; (d) posttranslational modifications of nascent α chains; (e) assembly and folding of three α chains into a triple helical molecule; (f) the stability of the helix structure, and (g) formation of cross-links between individual triple helical molecules. Large gene rearrangements can result in total loss of expression of the gene and small mutations in regulatory elements can have the same effects. Large and small mutations can also affect splicing of the primary transcript. Abnormal splicing and skipping of exons can cause two types of changes. Firstly, if the exon skipping is in frame, the result is a shortened polypeptide so that it lacks a segment and, secondly, if the exon skipping occurs out of frame, the result is a protein partially containing a nonsense amino acid sequence. Duplications of the gene segments can lead to abnormally long α chains.

Table 1. Single base mutations in COL4A5 in Alport syndrome

Mutation	Nucleotide change	Exon	RFLP	Phenotype							Ref.
				juvenile	adult	hearing loss	eye lesions	GBM splitting	kidney transplant	anti-GBM nephritis	
Missense											
Gly325→Arg	G→A at 1176[1]	17	−MspI		+	+	−	+	−		25
Gly325→Glu	G→A at 1177	17	−MspI			−	−	+			28
Gly521→Cys	G→T at 1764	23	+HindIII	+		+	−	nd	+	−	23
Gly1143→Asp	G→A at 3632	38	−MspI	+		+	+	+	+		30
Gly1143→Ser	G→A at 3630	38	−MspI		+	+	+/−	+	−		53
Pro1517→Thr	C→A at 3638	48		+		+	−	nd	−		31
Trp1538→Ser	G→C at 4702	48			+	−	−	+	+	−	27
Arg1563→Gln	G→A at 4777	48		+		+	+/−	+	+	−	31, 33
Cys1564→Ser	G→C at 4780	49	+PstI		+	+	−	+	+		18, 19
M1601→Ile	G→C at 4892	49		+		−	−	+			35
Nonsense											
Arg1563-Stop	C→T at 4776	48		+		+	−	nd			33
Splicing											
	G→C at 3657+1[2]	intron 38				+	−	+			32
	G→A at 3942−1	intron 41		+		+	−	+		−	31

[1] Numbering in cDNA sequence according to Zhou et al. [23].

[2] Numbering of intron mutations according to Beaudet and Tsui [52] refers to coding sequence in the first base of intron 38 following exon 38.

Table 2. Major rearrangements in COL4A5 in Alport syndrome

Mutation	Phenotype							Ref.
	juvenile	adult	hearing loss	eye lesions	GBM splitting	kidney transplant	anti-GBM nephritis	
Deletions								
Exons 42–47		+	+	−	nd	−		18
13 kb from 3' end of gene	+		+	−	+	+	−	22
Exon 1 extending upstream	+		+	+	+	−		24
Exon 1 extending upstream	+		−	+	+	+		24
Several 5' end exons and upstream	+		−	+	+	−		24
Several 5' end exons and upstream	+		−	+	+	−		58
>38 kb in 5' end	+		+	−	nd			29
Large deletion in 3' end	+		+	−	+	+	+	27
34 kb, exons 38–51	+		+	+	nd			26
Entire gene			+	+	nd	+	+	26
Deletion of GGGTGAA in exon 39	+		+	−	+	+		49
Deletion of TGGA in exon 41	+		+	+	+	+		49
Deletion of G in exon 50	+		+	−	+	+		49
Deletion of G in exon 39	+		+	−	+	−		54
Deletion of C in exon 42	+		+	nd	nd	+	−	54
Deletion of A in exon 34	+		+	+	+	+	−	55
Entire gene			+	−	nd	+	nd	56
Exon 1 and upstream			+	−	nd	nd	nd	56
Intragene deletion			+	+	+	+	nd	56
Intragene deletion			+	+	+	−	nd	56
Exon 17			+	nd	nd	+	nd	56
Exons 2–36	+		−	−	+	−	−	56
Exons 3–40/41			+	+	+	+	nd	56
Exons 39–49			+	+	nd	+	nd	56
Exons 38–41			+	−	nd	+	nd	56
Exons 38–39			−	+	+	−	−	56
Insertion/duplication/deletion								
Part of 3' end of gene			+	+	+	+		22
Insertion of TCCT in exon 45			−	−	+	−		57
Duplication/inversion								
Part of 3' end of gene			+	+	+	+	−	22
10 bp in exon 49		+	+	−	+			31

All the mutations leading to the absence of the α5(IV) chain or extensive alterations in protein size would obviously lead to abnormal structure of any type IV collagen network where the α5(IV) chain is required as a normal component. Complete loss or presence of an abnormal α5(IV) chain may explain the clinical manifestations observed in Alport syndrome. In the kidney the α5(IV) chain is normally present only in the GBM as shown by immunostaining techniques (see chapter 7). Very similar GBM staining is observed for the α3(IV) and α4(IV) chains (see chapter 8). Although it is not known in what type of chain combinations the α5(IV) chain is in type IV collagen trimer molecules, its strict GBM location indicates a specific role for the structure and function of the GBM. Consequently, the absence of a normal α5(IV) chain may weaken the structural network of the GBM in one way or another, leading to leakage of large proteins or even large blood cells into the urinary space. It is important to note, however, that heterozygous females usually exhibit only mild phenotypes and they seldom develop end-stage renal disease. This suggests that the synthesis of low amounts of a normal α5(IV) chain is sufficient to maintain normal GBM function, even though the abnormal chain is present. This fact is particularly important when considering future possibilities for gene therapy which may be accomplished in the future through the supplementation of expression of a normal transgene in addition to that of the mutated gene.

Single base mutations in exons leading to amino acid changes in the α5(IV) chain can be expected to render the protein malfunctional in most, but not necessarily all cases, since such changes can affect proteins in a number of ways. So, how can we know if an amino acid substitution actually causes disease instead of being silent? This is an important question, especially from a clinical point of view when making decisions about the termination of pregnancy after a mutation in the fetus has been identified. Actually, this is also a very difficult question to answer and, in fact, the only definite way of proving that a mutation is causative for the disease is to generate transgenic animals with the same mutation and observe if it results in Alport syndrome-like phenotype. However, it is possible that a mutation causing disease in man does not cause the same disease in mice. We, therefore, have to rely on strong 'circumstantial' evidence when predicting the potential pathogenic effect of mutations. Such evidence can be linkage or segregation of the mutations with the phenotype in an Alport kindred, knowledge about the role of certain amino acids for the function of the protein, or conservation of the amino acids during evolution. It is assumed that amino acids evolutionarily conserved in a protein between widely distant species have such functional importance that their substitution cannot be tolerated. This can be illustrated by several examples.

The about 230-residue amino acid sequence of the C-terminal NC1 domain of the type IV collagen α chains contains two homological halves (fig. 2) with about 50% of amino acids conserved in all six mammalian chains and also in type IV collagen α chains of *Drosophila, Caenorhabditis elegans* and sea urchin [7, 8, 16, 40–47, and X. Zhang et al., unpubl. results]. The conserved amino acids include twelve cysteines, six in each homologous half of the NC1 domain. It is apparent that the NC1 domain of all type IV collagen chains has the same biological function and that the highly conserved amino acids and three-dimensional structure of the NC1 domain are essential for the correct alignment of three α chains prior to their folding and twisting into a triple helix, and subsequently for the formation of intermolecular cross-links in the extracellular space (fig. 2). Therefore, substitution of conserved amino acids in this domain may affect these functions. Thus far, five single base mutations leading to amino acid substitutions in the NC1 domain of the α5(IV) chain have been published in Alport syndrome (fig. 3, table 1). One mutation resulted in the change of cysteine 1564 to serine [19]. The mutation would abolish one of the intrachain disulfide bonds, either destroying bonds between cysteines 1476 and 1564, or between cysteines 1508 and 1564 (fig. 4). The mutation which alters the conformation of the NC1 domain can have several effects. One possibility is that the mutation interferes with the alignment of chains and formation of the triple helix so that helix formation is slowed down, abolished completely, or the alignment of Gly-X-Y repeats is out of register. Alternatively, the triple helical molecule may be formed and the protein secreted from the cell, but the loss of the cysteine can cause weak intermolecular cross-links of the type IV collagen in the basement membrane meshwork.

The effect of amino acid changes in the NC1 domain other than those of cysteines are more difficult to predict as the function of those amino acids is not known. Examples for such mutations are the mutations proline 1517→threonine, tryptophan 1538→serine and arginine 1563→glutamine (table 1, fig. 3) [27, 31]. In those cases one can, if possible, (a) study if the mutation cosegregates in the families with the Alport phenotype, (b) see if the mutation is the result of a normal polymorphism and exists in normal individuals or, (c) one can examine if the amino acids are conserved during evolution. As can be seen in figure 5, all the amino acids listed above are conserved in all known type IV collagen α chains even between distant species. Consequently, it can be assumed that these amino acids are essential for the normal function of the NC1 domain and that their changes cannot be tolerated.

Several mutations converting a glycine residue in the Gly-X-Y repeat containing collagenous domain to another amino acid have been published (fig. 3, table 1). These mutations can without doubt affect the stability of the triple helix of the molecule because, as mentioned above, glycine is the only

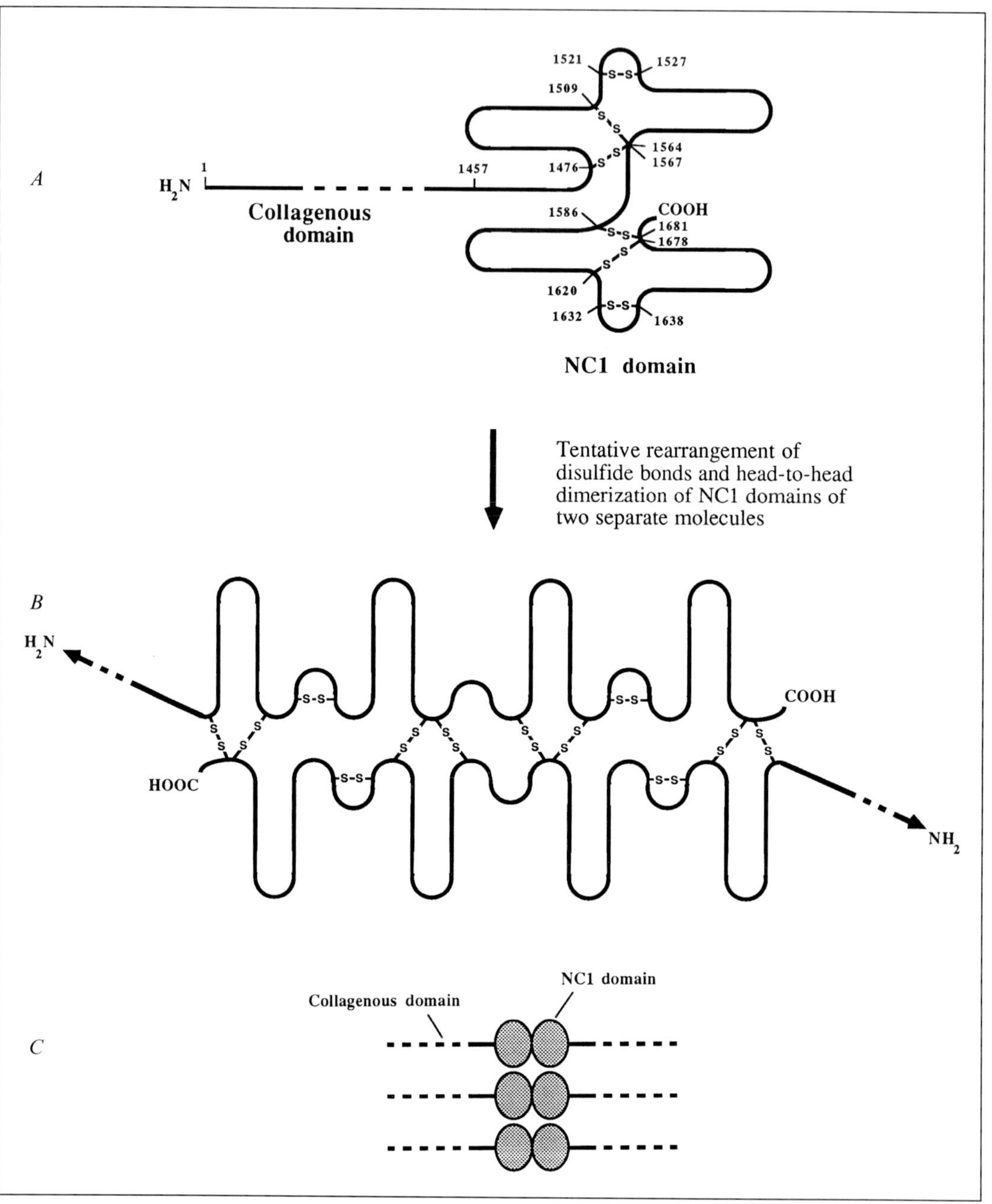

Fig. 2. Illlustration of normal intrachain disulfide bonds in the NC1 domain of a type IV collagen α1 chain (*A*), potential rearrangement of interchain disulfide bonds after dimerization of chains from two separate triple helical molecules via their NC1 domains

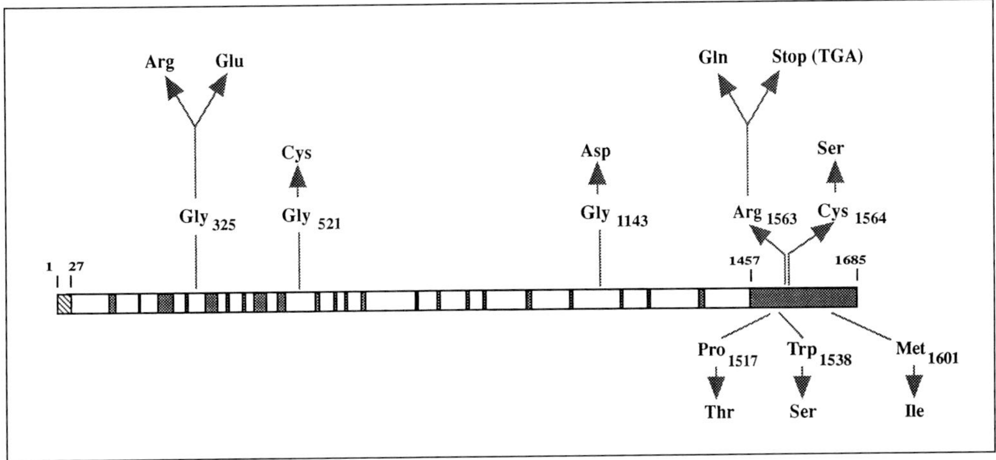

Fig. 3. Illustration of amino acid substitutions in the α5(IV) collagen chain resulting from single base mutations in the COL4A5 gene (see table 1). The signal peptide (residues 1–26) is depicted by cross-hatched pattern, the collagenous domain (residues 27–1456) by a white pattern interrupted by short noncollagenous sequences (shaded boxes), and the NC1 domain (residues 1457–1685) is shown by a shaded pattern.

amino acid small enough to suit into the center of the triple helix and an uninterrupted Gly-X-Y repeat sequence is essential for the maintenance of the triple helical conformation. The result of glycine substitution is, therefore, that the helix is destabilized, creating kinks in the molecule which are not tolerated. Although this has not been demonstrated at the protein level for type IV collagen, a number of similar glycine mutations, either inhibiting helix formation or causing kinks in the molecule, have been described in type I collagen in osteogenesis imperfecta [39].

The dispersion of mutations in the COL4A5 gene is a characteristic of Alport syndrome. Thus far, no 'hot spots' particularly prone to mutagenic changes have been found. However, out of the twelve single base mutations published thus far in Alport syndrome, two codons (Gly325 and Arg1563)

(*B*). For simplification, the bonds are shown only for single chains of the triple helical molecule. The NC1 domain, starting at amino acid residue 1457, contains two symmetrical halves. The positions of cysteine residues are numbered from the N-terminus of the protein according to Zhou et al. [23]. The exact organization of the disulfide bond between pairs 1476/1509 and 1564/1567 and pairs 1586/1620 and 1678/1681 has not been elucidated [modified from 19]. (*C*) All three NC1 domains of one triple helical type IV collagen molecule associate with an NC1 domain of another molecule in the dimerization (see chapter 3).

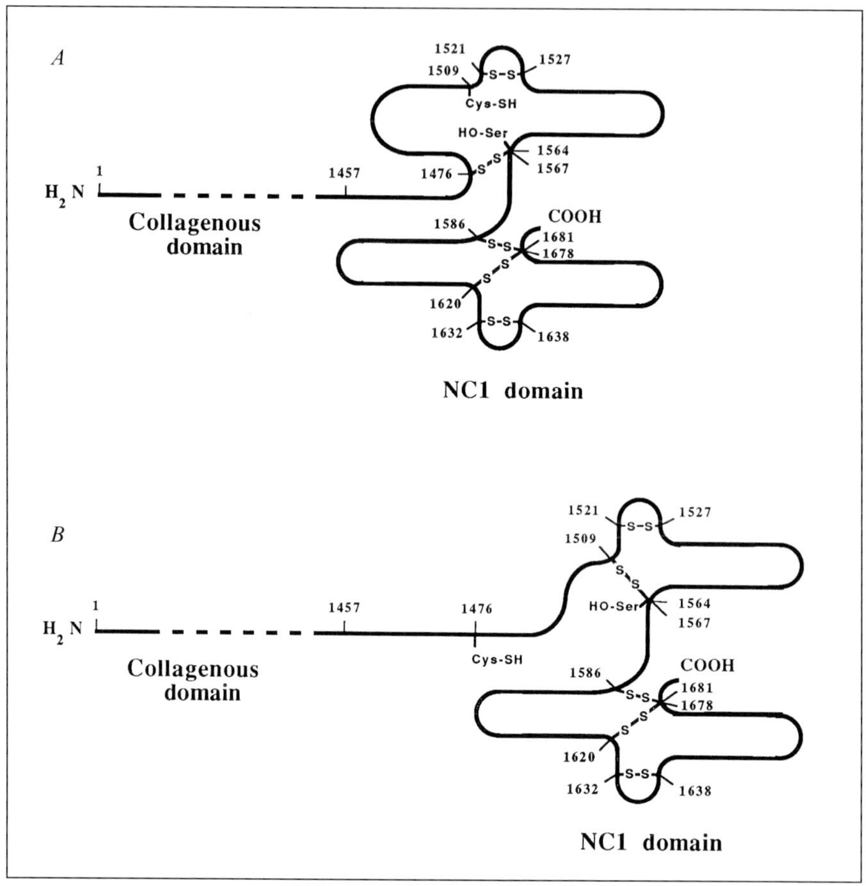

Fig. 4. Hypothetical consequences of a cysteine 1564→serine mutation in the COL4A5 gene. The mutation abolishes a disulfide bond connecting cysteine 1564 to either cysteine 1509 (*A*) or cysteine 1476 (*B*) [48]. This, in turn, causes a change in the conformation of the NC1 domain with possible rearrangement of intrachain bonds and subsequent interference of helix formation. The mutation may also seriously influence the rearrangement of the disulfide bonds and formation of interchain bonds, with the simplest change reducing the number of cross-links from 8 to 6. However, the mutation might interfere more seriously with dimerization by even completely inhibiting the formation of interchain cross-links between NC1 domains [modified from 19].

have been mutated in six apparently unrelated families [25, 28, 31, 33]. It remains to be seen whether or not these exons, or these particular codons, represent real 'hot spots'. Such regions which are highly susceptible to mutations would be of significant clinical value because they could greatly facilitate the search of mutations in Alport patients.

COL4A5 Mutations Cause Different Phenotypes

Alport syndrome is a heterogeneous disease with onset of end-stage renal disease at varying age and with or without manifestations, such as hearing loss and deafness, ocular lesions, thrombocytopathy, etc. Since so many mutations have been identified in the COL4A5 gene, it is of interest from a clinical point of view to correlate them with the phenotypes they produce. As can be seen from tables 1 and 2 and references therein, this analysis is disappointing in the sense that it does not seem to be possible to predict the phenotype based on the type of mutation. For example, the point mutations generated by substitution of glycine residues in the collagenous domain or conserved amino acids in the NC1 domain result in quite wide spectra of phenotypes (table 1). Most, but not all, patients have juvenile onset of end-stage renal disease, and the presence of hearing loss and ocular lesions varies between individuals.

A similar heterogeneous picture is seen for large changes in the gene (table 2). Even the entire loss of the gene does not necessarily produce more severe disease than a single glycine substitution in the collagenous domain of the $\alpha5(IV)$ chain. Although disappointing, this does not come as a surprise as the same phenomenon has been observed for defects in the genes for fibrillar collagens causing osteogenesis imperfecta, chondrodysplasia, Stickler syndrome and Ehlers-Danlos syndrome [39]. One particularly interesting Alport phenotype displays diffuse leiomyomatosis in addition to the classical symptoms [24]. The interesting molecular bases for this syndrome is discussed in detail in chapter 10. A clinically important group are the $\sim 15\%$ of posttransplantation Alport patients who develop anti-GBM antibodies. About one-third of those patients reject the transplant. Initial studies on Alport mutations and phenotypes gave hope that, based on the nature of mutation, it might be possible to predict which patients develop anti-GBM nephritis and, thus, who should obtain a transplant and who not. It seemed reasonable that a mutation deleting the highly antigenic NC1 domain of the $\alpha5(IV)$ chain would cause the patient to recognize the corresponding domain in the allograft as a foreign protein and should, therefore, not receive an allograft as he would develop anti-GBM nephritis. however, more recent data on transplanted patients with similar kinds of gene defects (table 2) have shown that this explanation is not totally valid. In the light of the current and still limited knowledge, it is not yet possible to predict which patients are going to develop anti-GBM nephritis following kidney transplantation.

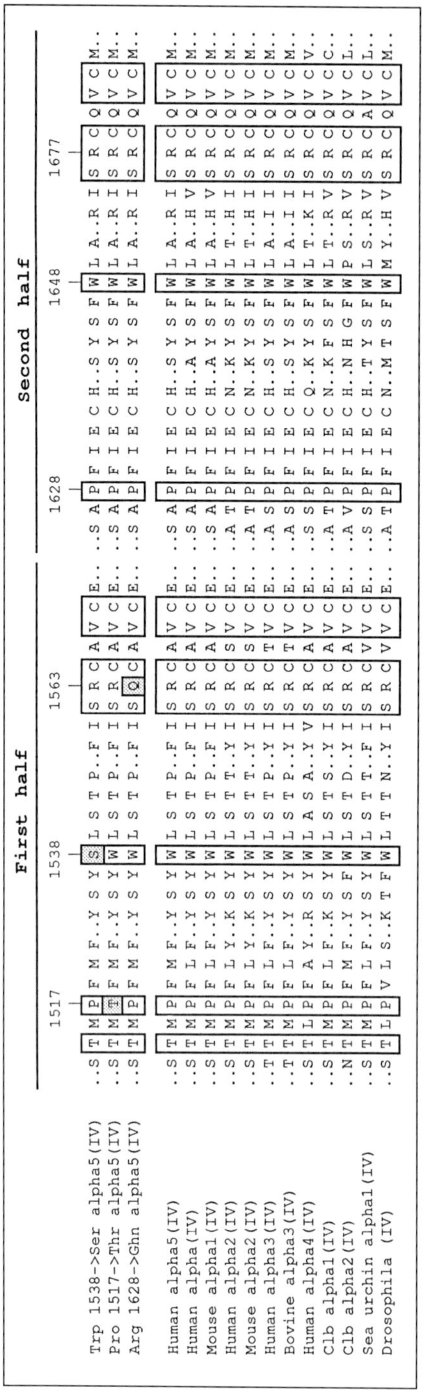

Fig. 5. Comparison of substituted amino acids in three Alport patients with corresponding residues in the both homologous halves of the NC1 domain in all known type IV collagen α chains [7, 8, 16, 40–47, and X. Zhang et al., unpubl. results]. Gaps (..) have been introduced to maintain alignment. Capital letters indicate the amino acid residues. Amino acids that are conserved through evolution are shown in boxes. The proline, threonine, and arginine (boldface) which are mutated in the three patients [31] are conserved in both symmetrical halves of the NC1 domain in all known type IV α chains.

Involvement of COL4A5, COL4A4 and COL4A6 Collagen Genes in Alport Syndrome

Although Alport syndrome is primarily considered to be an X-chromosome-linked disease, autosomal forms have also been postulated based on pedigree data. After defects were demonstrated in the α5(IV) collagen gene in the X-linked disease, it was logical to search for mutations in the COL4A3 and COL4A4 genes for the α3(IV) and α4(IV) collagen chains in the autosomal form of the disease, respectively, as these chains are present in the GBM. Recently, Reeders' group [20] identified the first mutations in both the COL4A3 and COL4A4 genes in four Alport kindreds. All the patients were homozygous for their mutations, while heterozygous parents or siblings had only mild findings or were unaffected. One mutation in COL4A3 was found in a girl with juvenile Alport accompanied by sensorineural deafness. Her brother has similar disease, but the parents are unaffected. The girl was transplanted at the age of 10, but developed anti-GBM nephritis. The mutation is a 5-bp deletion in exon 47 leading to a premature stop codon 34 or 35 amino acids downstream. Another COL4A3 mutation was found in a girl who had proteinuria and microscopic hematuria at age 7, leading to end-stage renal disease and transplantation at age 11. She had sensorineural hearing loss and lamellation and splitting of the GBM. The parents are normal, although the father has mild persistent microscopic hematuria. The mutation is a single base substitution leading to the conversion of the codon for arginine 1481 in the NC1 domain to a stop codon. Two mutations were also described in COL4A4 [20]. One mutation found in 2 young sisters was a single nucleotide change causing a glycine 1201→serine substitution. One of the girls had GBM changes at age 7 and end-stage renal disease at age 14. Her sister had hematuria at age 5 and end-stage renal disease at age 11. No hearing or ocular abnormalities were apparent. The parents and 1 sister have no renal manifestations. The second COL4A4 mutation was found in 3 young sisters with progressive renal disease without hearing loss. The mutation replaced the codon for a serine in the collagenous domain with stop codon causing premature chain termination and shortening of the chain by 453 amino acids. Based on those findings, it is apparent that homozygous mutations in the genes for the α3(IV) and α4(IV) collagen chains lead to similar Alport syndrome phenotypes as hemizygous mutations in the X-linked form, where the gene for the α5(IV) chain is affected. This indicates that these three chains serve similar functions, possibly by being present in the same heterotrimers. However, details of the arrangement of chain assembly in type IV collagen molecules of the GBM are still not fully understood.

At the present there is no direct evidence reported showing that mutations in the α6(IV) collagen chain COL4A6 gene [51] alone can cause Alport syndrome. Deletions in the 5' end of the COL4A6 gene extending into the adjacent COL4A5 gene have been reported to cause Alport syndrome and diffuse leiomyomatosis (see chapter 9). In our preliminary work we have sequenced four exons in the 5' end of the COL4A6 gene in over 200 patients without finding any mutation [Heiskari et al., unpubl. data]. Considering the random localization of mutations in the COL4A5 gene in Alport syndrome, it was assumed that the same would hold for COL4A6 if it were involved in the disease. These initial data indicate that (a) either the COL4A6 gene is not mutated in Alport syndrome, or (b) the mutations are not located randomly in the gene. Present data support the former hypothesis as the COL4A6 gene appears to differ from COL4A5 with respect to spatial expression [50]. However, exclusion of COL4A6 as being mutated in some Alport kindreds remains to be verified.

References

1   Kinoshita Y, Osawa G, Morita T, Kobayashi N, Wada J, Ebe T, Watanabe M, Muronashi K, Muryayama M: Hereditary chronic nephritis (Alport) complicated by nephrotic syndrome. Light, fluorescent and electron microscopic studies of renal biopsy specimens. Acta Med Biol Niigata 1969;17:101–117.

2   Hinglais N, Grünfeld J-P, Bois E: Characteristic ultrastructural lesions of the glomerular basement membrane in progressive hereditary nephritis (Alport's syndrome). Lab Invest 1972;27:473–487.

3   Spear GS, Slusser RJ: Alport's syndrome: Emphasizing electron microscopic studies of the glomerulus. Am J Pathol 1972;69:213–224.

4   Churg J, Sherman RL: Pathological characteristics of hereditary nephritis. Arch Pathol 1973;95:374–379.

5   Kefalides NA: Isolation of a collagen from basement membranes containing three identical α-chains. Biochem Biophys Res Commun 1971;45:226–234.

6   Spear GS: Alport's syndrome: A consideration of pathogenesis. Clin Nephrol 1973;1:336–337.

7   Pihlajaniemi T, Tryggvason K, Myers J, Kurkinen M, Lebo R, Cheung M, Prockop DJ, Boyd CD: cDNA clones for the pro-α1(IV) chain of human type IV procollagen reveal an unusual homology of amino acid sequences in two halves of the carboxyl-terminal domain. J Biol Chem 1985;260:7681–7687.

8   Hostikka SL, Kurkinen M, Tryggvason K: Nucleotide sequence coding for the human type IV collagen α2 chain cDNA reveals extensive homology with the NC-1 domain of α1(IV), but not with the collagenous domain or untranslated region. FEBS Lett 1987;216:281–286.

9   Griffin C-A, Emanuel BS, Hansen JR, Cavenee WK, Myers JC: Human collagen genes encoding basement membrane α1(IV) and α2(IV) chains map to the distal long arm of chromosome 13. Proc Natl Acad Sci USA 1987;84:512–516.

10 Butkowski RJ, Langeveld JPM, Wieslander J, Hamilton J, Hudson BG: Localization of the Goodpasture epitope to a novel chain of basement membrane collagen. J Biol Chem 1987;262:7874–7877.

11 Saus J, Wieslander J, Langeveld J, Quinones S, Hudson BG: Identification of the Goodpasture antigen as the α3(IV) chain of collagen IV. J Biol Chem 1988;263:13374–13380.

12   Kleppel MM, Kashtan CE, Butkowski RJ, et al: Alport familial nephritis: Absence of 28 kilodalton non-collagenous monomers of type IV collagen in glomerular basement membrane. J Clin Invest 1987;80:263–266.
13   Atkin CL, Hasstedt SJ, Menlove L, Cannon L, Kirchner N, Schwartz N, Nguyen K, Skolnick M: Mapping of Alport syndrome to the long arm of the X-chromosome. Am J Hum Genet 1988;42: 249–255.
14   Brunner H, Schröder C, van Bennekom C, Lambermon E, Tuerlings J, Menzel D, Olbing H, Monnens L, Wieringa B, Robers HH: Localization of the gene for X-linked Alport's syndrome. Kidney Int 1988;34:507–510.
15   Flinter FA, Abbs S, Bobrow M: Localization of the gene for classic Alport syndrome. Genomics 1989;4:335–338.
16   Hostikka SL, Eddy RL, Byers MG, Höyhtyä M, Shows TB, Tryggvason K: Identification of a distinct type IV collagen α chain with restricted kidney distribution and assignment of the gene to the locus of X-chromosome-linked Alport syndrome. Proc Natl Acad Sci USA 1990;87:1606– 1610.
17   Pihlajaniemi T, Pohjalainen E-R, Myers J: Complete primary structure of the triple helical region and the carboxy-terminal domain of a new type IV collagen chain $\alpha5(IV)$. J Biol Chem 1990;265: 13758–13766.
18   Barker D, Hostikka SL, Zhou J, Chow LT, Oliphant AR, Gerken SC, Gregory MC, Skolnick MH, Atkin CL, Tryggvason K: Identification of mutations in the COL4A5 collagen gene in Alport syndrome. Science 1990;248:1224–1227.
19   Zhou J, Barker D, Hostikka SL, Gregory M, Atkin C, Tryggvason K: Single base mutation in $\alpha5(IV)$ collagen chain gene converting a conserved cysteine to serine in Alport syndrome. Genomics 1991;9:10–18.
20   Mochizyki T, Lemmink H, Mariyama M, Antignac C, Gubler M-C, Pirson Y, Verellen-Dumoulin C, Chan B, Schröder C, Smeets HJM, Reeders ST: Identification of mutations in the $\alpha3(IV)$ and $\alpha4(IV)$ collagen genes in autosomal recessive Alport syndrome. Nat Genet 1992;8:77–82.
21   Kivirikko K, Myllylä R: Recent developments in posttranslational modification: Intracellular processing. Methods Enzymol 1987;144:96–114.
22   Boye E, Vetrie D, Flinter F, Buckle B, Pihlajaniemi T, Hämäläinen ER, Myers JC, Bobrow M, Harris A: Major rearrangements in the $\alpha5(IV)$ collagen gene in three patients with Alport syndrome. Genomics 1991;11:1125–1132.
23   Zhou J, Herz JM, Leinonen A, Tryggvason K: Complete amino acid sequence of the human $\alpha5$ collagen chain and identification of a single base mutation in exon 29 from the 3' end converting glycine-521 in the collagenous domain to cysteine in an Alport syndrome patient. J Biol Chem 1992;267:12475–12481.
24   Antignac C, Zhou J, Sanak M, Cochat P, Rousssel B, Deschênes G, Knebelmann B, Hors MC, Tryggvason K, Gubler MC: Alport syndrome and diffuse leiomyomatosis – Deletions in the 5' end of the COL4A5 collagen gene. Kidney Int 1992;42:1178–1183.
25   Knebelmann B, Deschênes G, Gros F, Hors MC, Grünfeld J-P, Zhou J, Tryggvason K, Gubler M-C, Antignac C: Substitution of arginine for glycine-325 in the collagen $\alpha5(IV)$ chain associated with X-linked Alport syndrome: Characterization of the mutation by direct sequencing of PCR-amplified lymphoblast cDNA fragments. Am J Hum Genet 1992;51:135–142.
26   Netzer KO, Renders L, Zhou J, Pullig O, Tryggvason K, Weber M: Deletions of the COL4A5 gene in patients with Alport syndrome. Kidney Int 1992;43:1336–1344.
27   Smeets HJM, Melenhorst JJ, Lemmink HH, Schröder CH, Nelen MR, Zhou J, Hostikka SL, Tryggvason K, Ropers HH, Jansaweijer MCE, Monnens LAH, Brunner HG, van Oost BA: Different mutations in the COL4A5 collagen gene in two patients with different features of Alport syndrome. Kidney Int 1992;42:83–88.
28   Renieri A, Seri M, Myers JC, Pihlajaniemi T, Massella L, Rizzoni G, De Marchi M: De novo mutation in the COL4A5 gene converting glycine-325 to glutamic acid in Alport syndrome. Hum Mol Genet 1992;1:127–129.
29   Renieri A, Seri M, Myers JC, Pihlajaniemi T, Sessa A, Rizzoni G, De Marchi M: Alport syndrome caused by a 5' deletion within the COL4A5 gene. Hum Genet 1992;89:120–121.

30 Zhou J, Hertz JM, Tryggvason K: Mutation in the α5(IV) collagen chain in juvenile-onset Alport syndrome without hearing loss or ocular lesions: Detection by denaturing gradient gel electrophoresis of a PCR product. Am J Hum Genet 1992;50:1291–1300.

31 Lemmink HH, Schröder CH, Brunner HG, Nelen MR, Zhou J, Tryggvason K, Haagsma-Schouten WAG, Roodvoets AP, Rascher W, van Oost BA, Smeets JM: Systematic screening for mutations in 16 exons of the COL4A5 in Alport syndrome. Genomics 1993;17:485–489.

32 Netzer KO, Pullig O, Zhou J, Frei U, Tryggvason K, Weber M: COL4A5 splice site mutation and α5(IV) collagen mRNA in Alport syndrome. Kidney Int 1993;43:486–492.

33 Zhou J, Gregory MC, Hertz JM, Barker DF, Atkin C, Spencer E, Tryggvason K: Mutations in the codon for a conserved arginine-1563 in the COL4A5 collagen gene in Alport syndrome. Kidney Int 1993;43:722–729.

34 Boye E, Flinter FA, Bobrow M, Harris A: An 8 bp deletion in exon 51 of the COL4A5 gene of an Alport syndrome patient. Hum Mol Genet 1993;2:595–596.

35 Nomura S, Osawa G, Sai T, Harano T, Harano K: A splicing mutation in the alpha 5(IV) collagen gene of a family with Alport's syndrome. Kidney Int 1993;43:116–1124.

36 Zhou J, Leinonen A, Tryggvason K: Structure of the human type IV collagen COL4A5 gene. J Biol Chem 1994;269:6608–6614.

37 Vetrie D, Flinter F, Bobrow M, Harris A: Long-range mapping of the gene for the human alpha 5(IV) collagen chain at Xq22–q23. Genomics 1992;12:130–138.

38 Kerem BS, Rommens JM, Buchanan JA, Markiewicz D, Cox TK, Chakravarti A, Buchwald M, Tsui LC: Identification of the cystic fibrosis gene: Genetic analysis. Science 1989;245:1073–1080.

39 Kuivaniemi H, Tromp F, Prockop DJ: Mutations in collagen genes: Causes of rare and some common diseases in humans. FASEB J 1991;5:2052–2060.

40 Hostikka SL, Tryggvason K: The complete primary structure of the α2 chain of human type IV collagen and comparison with the α1(IV) chain. J Biol Chem 1988;263:19488–19493.

41 Blumberg B, MacKrell AJ, Fessler JH: Drosophila basement membrane procollagen alpha 1(IV). II. Complete cDNA sequence, genomic structure and general implications for supramolecular assemblies. J Biol Chem 1988;263:18328–18337.

42 Guo X, Kramer JM: The two *Caenorhabditis elegans* basement membrane (type IV) collagen genes are located on separate chromosomes. J Biol Chem 1989;264:17574–17582.

43 Morrison KE, Mariyama M, Yang-Feng TL, Reeders ST: Sequence and localization of a partial cDNA encoding the human α3 chain of type IV collagen. Am J Hum Genet 1991;49:545–554.

44 Muthukumaran G, Blumberg B, Kurkinen M: The complete primary structure for the α1-chain of mouse collagen IV. Differential evolution of collagen IV domains. J Biol Chem 1989;264:6310–6317.

45 Exposito JY, D'Alessio M, Di Leberto M, Ramirez F: Complete primary structure of a sea urchin type IV collagen α chain and analysis of the 5' end of its gene. J Biol Chem 1993;268:5249–5254.

46 Mariyama M, Kalluri R, Hudson BG, Reeders ST: The α4(IV) chain of basement membrane collagen. Isolation of cDNAs encoding bovine α4(IV) and comparison with other type IV collagens. J Biol Chem 1992;267:1253–1258.

47 Saus J, Quinones S, MacKrell A, Blumberg B, Muthukumaran G, Pihlajaniemi T, Kurkinen M: The complete primary structure of mouse α2(IV) collagen. Alignment with mouse α1(IV) collagen. J Biol Chem 1989;264:6318–6324.

48 Siebold B, Deutzmann R, Kühn K: The arrangement of intra- and intermolecular disulfide bonds in the carboxyterminal, non-collagenous aggregation and cross-linking domain of basement-membrane type IV collagen. Eur J Biochem 1988;176:617–624.

49 Renieri A, Seri M, Galli L, Cosci P, Imbasciati E, Massela L, Rizzoni G, Restagno G, Carbonara AO, Stramignoni E, Basolo B, Piccoli G, De Marchi M: Small frameshift deletions within the COL4A5 gene in juvenile-onset Alport syndrome. Hum Genet 1993;92:417–420.

50 Hudson BG, Reeders ST, Tryggvason K: Type IV collagen: Structure, gene organization, and role in human diseases: Molecular basis of Goodpasture and Alport syndromes and diffuse leiomyomatosis. Minireview. J Biol Chem 1993;268:26033–26036.

51 Zhou J, Mochizuki T, Smeets H, Antignac C, Laurila P, de Paepe A, Tryggvason K, Reeders ST: Deletion of the paired α5(IV) and α6(IV) collagen genes in inherited smooth muscle tumors. Science 1993;261:1167–1169.

52 Beaudet AL, Tsui LC: A suggested nomenclature for designating mutations. Hum Mutat 1993;2: 245–248.

53 Renieri A, Meroni M, Sessa A, Battini G, Serbelloni P, Torri Tarelli L, Seri M, Galli L, De Marchi M: Variability of clinical phenotype in a large Alport family with gly1143→ser change of collagen α5(IV) chain. Nephron 1994;67:444–449.

54 Renieri A, Galli L, De Marchi M, Li Volti S, Mollica F, Lupo A, Maschio, Peissel B, Rossetti S, Pignatti, Turco A: Single base pair deletions in exons 39 and 42 of the COL4A5 gene in Alport syndrome. Hum Mol Genet 1994;3:201–202.

55 Peissel B, Rossetti S, Renieri A, Galli L, De Marchi M: A novel frameshift deletion in type IV collagen α5 gene in a juvenile-type Alport syndrome patient. Hum Mutat 1994;3:386–390.

56 Antignac C, Knebelmann B, Drouot L, Deschênes G, Hors-Cayla M-C, Zhou J, Tryggvason K, Grünfeld J-P, Broyer M, Gubler M-C: Deletions in the COL4A5 collagen gene in X-linked Alport syndrome: Characterization of the pathological transcripts in non-renal cells and correlation with disease expression. J Clin Invest 1994;93:1195–1207.

57 Massella L, Rizzoni G, De Blasis R, Barsotti P, Faraggiana T, Renieri A, Seri M, Galli L, De Marchi M: De novo COL4A5 gene mutations in Alport syndrome. Nephrol Dial Transplant 1994; 9:1408–1411.

58 Renieri A, Bassi MT, Galli L, Zhou J, Giani M, De Marchi M, Ballabioi A: A deletion spanning the 5' ends of both the COL4A5 and COL4A6 genes in a patient with Alport syndrome and leiomyomatosis. Hum Mutat 1994;4:195–198.

Karl Tryggvason, MD, PhD, Department of Medical Biochemistry and Biophysics, Karolinska Institute, S-171 77 Stockholm (Sweden)

Tryggvason K (ed): Molecular Pathology and Genetics of Alport Syndrome.
Contrib Nephrol. Basel, Karger, 1996, vol 117, pp 172–182

Mutations in Alport Syndrome Associated with Diffuse Esophageal Leiomyomatosis

Corinne Antignac, Laurence Heidet[1]

INSERM U423, Hôpital Necker – Enfants Malades, Paris, France

Diffuse esophageal leiomyomatosis (DL) is a rare condition characterized by benign smooth muscle cell proliferation [1]. Sporadic and hereditary cases have been described [2–4]. Other organs, especially the female genital tract and the tracheobronchial tree, may also be involved. The association of DL with Alport syndrome (AS) was first stressed by Garcia-Torres and Guarner [5]. At least 35 cases from 19 families have now been documented, and cosegregation of AS and DL (DL-AS) has been observed in all male cases [5–20]. Therefore, the association of DL-AS cannot be considered as pure coincidence. In our first report concerning molecular anomalies in 3 male patients with DL-AS, we showed that the patients had a deletion including the 5′ end of the COL4A5 gene and extending beyond it [19]. More recently, the COL4A6 gene, encoding a novel collagen IV α6 chain, has been cloned and shown to be located 450 bp upstream of COL4A5 [20]. The deletion in COL4A6 is confined to the first two exons of this gene in these 3 patients as well as in 5 other DL-AS patients [20, 21]. However, 3 patients with AS but without DL and bearing a deletion of the 5′ part of COL4A5 have been shown to have larger deletion in COL4A6 [21]. Three different hypotheses could explain such results: (1) DL could be due to the absence of α6(IV) chain; (2) DL could involve a third gene located in the large second intron of COL4A6, or (3) DL could be due to the presence of an

[1] We thank Dr. C. Kashtan for helpful discussion and D. Grauz for aiding in the preparation of the manuscript. This work was supported by INSERM, AP-HP, AFM, CNAMS, MGEN, ARC and the Ligue Nationale contre le Cancer.

abnormal truncated α6(IV) chain. The fact that a COL4A6 mRNA product has been detected in an esophageal tumor sample of a patient with DL-AS highly favors this latter hypothesis [21].

Clinical and Pathological Features of the Syndrome

At least 35 cases belonging to 19 families have been described in the literature since the first cases described by Garcia-Torres and Guarner [5]. Furthermore, 2 cases previously described by Kenney [22] and Johnston et al. [23], probably belong to the same syndrome. If all the cases presented with DL, the leiomyomatosis process is not limited to the esophagus, and extends, in many cases, to the tracheobronchial tree and to the genital tract in females. Furthermore, a unique feature in this syndrome is the high incidence of severe, congenital and bilateral cataracts, a type of ocular involvement usually not observed in typical AS. The main clinical features of these cases are summarized in table 1. There are 14 males and 21 females.

Leiomyomatosis
Diffuse Esophageal Leiomyomatosis. The age of onset is frequently difficult to define precisely, since there is often a long history of digestive or respiratory symptoms before the first exam. However, the average age of onset of the symptoms appears to be 6 years in males and 10 years in females. Thus, DL seems to develop earlier among male patients. However, the symptoms can occur as early as 1 year, even in a female patient.

The clinical manifestations are often severe, and are either digestive or respiratory. The main symptoms are dysphagia, postprandial vomiting, retrosternal or epigastric pain, and recurrent bronchitis. Dyspnea, cough, stridor and apnea (in a 1-year-old child) have also been described. Bleeding is very unusual. Nearly or completely asymptomatic patients have also been described (patient 32).

If DL is suspected after chest radiograph or barium opacification of the esophagus, computed tomography or magnetic resonance imaging are valuable to confirm the diagnosis, by demonstrating a thickening of the esophageal wall. Actually, without these investigations, a percentage of DL could be misdiagnosed as extrinsic lesions or as an esophageal motility disorder (such as achalasia [9, 15]), because of the infiltration of the leiomyomatosis process into the myenteric plexus.

Macroscopically, a diffuse thickening of the esophageal wall is observed (fig. 1), whose extension is variable, mainly predominating in the lower third of the esophagus, but could extend to the upper third of the esophagus or

Case No.	Family No.	Age (years)[1] sex	Organs with leiomyo-mata	DL symptoms	age of onset, years	Nephro-pathy	Deaf-ness	Ocular symp-toms	Family history	Ref.
		17M	E	dysphagia	6	–	yes	bilat. cong. cataracts	–	22
		18F	E, G	dysphagia	15	hematuria	–	–	–	23
1	1	30F	E, G, TB	dysphagia retrosternal pain hematemesis	27	hematuria	–	–	+	5
2	1	20M	E, TB	dysphagia retrosternal pain	14	hematuria ESRD	yes	bilateral cataracts incipiens	+ son of case No. 1	5
3	1	17F	E, G	dysphagia retrosternal pain recurrent bronchitis, vomiting	< 12	hematuria	no	bilateral cataracts incipiens	+ daughter of case No. 1	5
4	1	7M	E	dysphagia	–	hematuria EM +	no	no	+ son of case No. 3	6
5	1	4M	E	dysphagia	–	hematuria EM +	yes	bilat. cong. cataracts	+ son of case No. 3	6
6	2	19F	E, G, TB	dysphagia vomiting	10	hematuria EM +	no	bilat. cong. cataracts	no	5
7	3	37F	E	intermittent dysphagia	15	hematuria	–	–	–	5
8	4	20M	E	vomiting retrosternal pain dyspnea	9	hematuria ESRD EM +	yes	bilat. cong. cataracts	+	7
9	4	– F	E, G	–	–	hematuria	no	no	+ mother of case No. 8	7
10	5	7M	E	dysphagia recurrent pulmonary infections	–	hematuria EM +	no	bilat. cong. cataracts	+	8
11	5	– F	E, G	–	–	hematuria	no	–	+ mother of case No. 10	8

Case No.	Family No.	Age (years)[1] sex	Organs with leiomyo-mata	DL symptoms	age of onset, years	Nephro-pathy	Deaf-ness	Ocular symp-toms	Family history	Ref.
12	6	8M	E	wheezing, dyspnea, post-prandial vomiting	5	hematuria CRF EM+	yes	bilateral lenticonus	no	9
13	7	9M	E	cough comiting	–	hematuria EM+	–	–	no	10
14	8	17F	E	recurrent bronchitis	6	no	no	bilat. cong. cataracts	no	11
15	9	39F	E, G, TB, P	dysphagia bronchospasm	20	no EM–	no	bilateral cataracts	+	11
16	9	11F	E	recurrent bron-chitis, vomiting painful dysphagia	2	hematuria	no	bilateral cataracts	+daughter of case No. 15	11
17	10	27F	E	painful dysphagia	–	–	–	–	+	11
18	10	9M	E	frequent vomiting	5	hematuria EM+	yes	bilateral cataracts	+son of case No. 17	11
19	11	26M	E	asthma	–	ESRD	yes	bilat. len-ticonus, macular flecks	+	12
20	11	–M	E	–	–	AS	–	–	+brother of case No. 19	12
21	11	–M	E	–	–	AS	–	–	+brother of case No. 19	12
22	11	–F	E	–	–	no symptom	–	–	+mother of case No. 19	12
23	12	26F	E, G, U	epigastric pain	4	hematuria	–	–	–	13
24	13	14M	E	dysphagia ret-rosternal pain pneumonia, food regurgitation with choking	4	hematuria EM+	no	bilat. cong. cataracts	no	14

Case No.	Family No.	Age (years)[1] sex	Organs with leiomyomata	DL symptoms	age of onset, years	Nephropathy	Deafness	Ocular symptoms	Family history	Ref.
25	14	13F	E, P	dysphagia postprandial vomiting	12	AS	–	–	–	15
26	15	5F	E, P	recurrent pulmonary infections	4	hematuria	no	bilat. cong. cataracts	no	16
27	16	5F	E	recurrent vomiting, epigastric pain	3	hematuria	no	no	+	17
28	16	3F	E	stridor, apnea recurrent bronchitis	1	hematuria	no	no	+ sister of case No. 27	17
29	16	26F	E	intermittent painless vomiting	–	hematuria	no	no	+ mother of case No. 27	17
30	16	– F	E	dysphagia regurgitation	8	–	no	no	+ mother of case No. 29	17
31	17	21F	E	dysphagia epigastric pain	–	hematuria	no	no	no	18
32	18	25M	E	routine chest radiography	–	hematuria ESRD EM+	no	anterior lenticonus	–	19
33	19	– F	E, G	–	–	hematuria	–	–	+	20
34	19	23M	E	dysphagia	–	hematuria ESRD	yes	bilateral cataracts	+ brother of case No. 33	20
35	19	– F	E	–	–	hematuria EM+	–	bilateral cataracts	+ daughter of case No. 33	20

E = Esophagus; G = genital involvement; P = perineum; U = urethra; TB = tracheobronchial involvement; EM+ = typical ultrastructural changes of AS; CRF = chronic renal failure; ESRD = end-stage renal disease; – = not mentioned or not studied.

[1] Age at time of report of last follow-up.

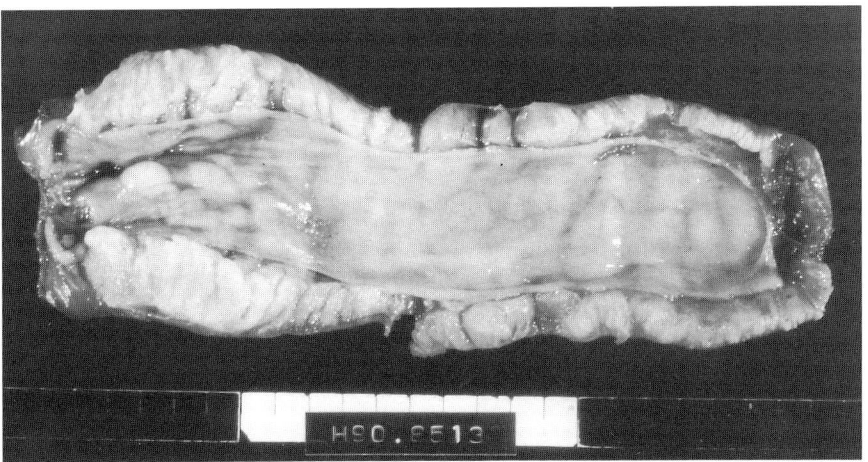

Fig. 1. Resection specimen of an esophagus with leiomyomatosis, showing the diffuse thickening of the esophageal wall (with the courtesy of P. Cochat).

of the stomach. Histologically, the thickened esophageal musculature shows extensive replacement of the normal fiber pattern by irregular, plexiform fibers with whorl formation. There are minimal atypia, no detectable mitoses and an overall low cellularity. DL may affect either one or both of the circular and longitudinal muscle layers of the esophagus. There is variable involvement of the muscularis mucosae. Some visible isolated nodules or even a confluent nodular pattern are sometimes present but always superimposed on a diffuse process. The DL lies beneath a normal esophageal and gastric epithelium. Other recorded features include prominence of the nerve plexi and a mild chronic inflammatory cell infiltrate. In most cases, surgical treatment is necessary and consists of partial or subtotal esophagectomy with or without partial gastrectomy and reconstruction by intestinal interposition.

Other Sites of Smooth Muscle Cell Proliferation. Tracheobronchial leiomyomatosis, with a pediculated leiomyoma of the carina in 1 case, has been documented in the 4 cases in which autopsy was performed. It was responsible for bronchospasm and death in 1 patient. Thus, the tracheobronchial involvement is probably very frequent, and might account for the episodes of paroxysmal dyspnea observed in some patients. Genital involvement was present in 9 of 14 adult females, mainly as clitoral hypertrophy, vulvar, perivaginal or uterine leiomyomas. Periurethral [13] and perirectal [11, 15, 16] leiomyomas have also been described.

Associated Cancers. It is noteworthy, even if it is not possible to link these different features, that 2 patients presented ovarian tumors (dysembryoma in case 15 and mesenchymal tumor in case 23), and that 2 young adult females (27 and 39 years old) died from unusual carcinomas (gallbladder and esophageal carcinomas, the latter arising in an endobrachyesophagus) (cases 15 and 17).

Alport Syndrome

All patients have typical clinical renal features and GBM ultrastructural changes (when available) of AS. In each family, the pedigree is compatible with an X-linked transmission, with no male-to-male transmission and more severe disease in males. Furthermore, all male patients have a severe phenotype. Indeed, 9 male patients have chronic renal failure or have reached ESRD around 20 (all but 1 of them are deaf), and 5 other 4- to 9-year-old male patients seem to have juvenile-type AS as suggested by the early onset of renal symptoms (hematuria discovered at birth in 2 children and before 3 in the others) and deafness (2 of them have already developed hearing loss). The female patients have milder symptoms. None of them developed renal failure. The only renal symptom reported is microscopic hematuria. Deafness is never reported. In 1 case (No. 15), a 39-year-old female patient with severe DL has no renal symptom and did not display any ultrastructural GBM changes, whereas her daughter has hematuria.

Ocular Symptoms

Typical AS ocular symptoms (i.e. lenticonus and/or macular flecks) are only reported in 3 patients. By contrast, congenital bilateral cataracts, which are not a feature of classical AS, have been reported in 7 male patients and in 6 female patients and, thus, occurred in at least one third of the patients. However, the penetrance of the disorder is not complete, not only in females, which is usual for an X-linked disorder, but also in males, since in a kindred with 2 male children affected with DL-AS, only 1 had cataracts [6].

Molecular Biology

The first molecular studies dealing with DL-AS [19, 20] have shown the presence of deletions removing the 5' end of both COL4A5 and COL4A6 genes in 4 unrelated DL-AS patients (3 males and 1 female) (fig. 2, 3). More recently, a molecular study of 4 additional DL-AS patients has shown the same type of rearrangement [21]. It is of interest to notice that, in all cases, the deletion involves only the two first exons of COL4A6, whereas its extent in COL4A5 is variable from one patient to another. Furthermore, 3 patients

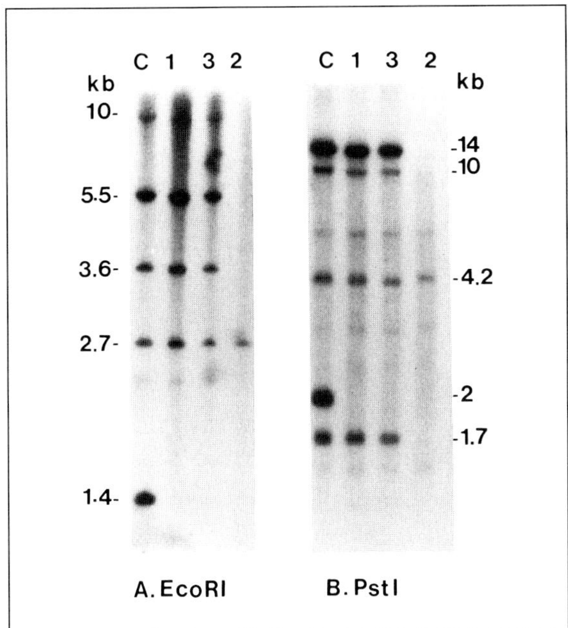

Fig. 2. Southern blots of EcoRI-digested DNAs from DL-AS patients, probed to the COL4A5 cDNA probe JZ4 [37]. All patient DNAs (lanes 1–3) lack the 1.4-kb fragment corresponding to COL4A5 exon 1 present in the control DNA (C). Patient 2 DNA also lacks the 10-, 5.5- and 3.6-kb fragments spanning COL4A5 exons 2–9.

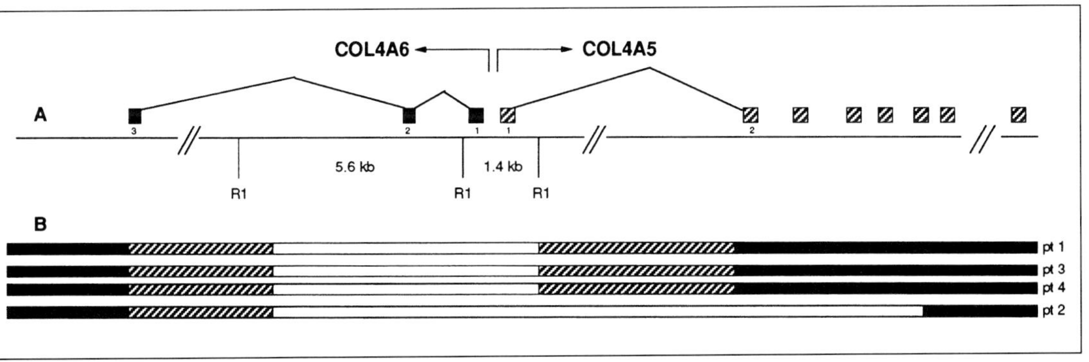

Fig. 3. Schematic diagram of the extent of the deletions in the COL4A5 and COL4A6 genes in 4 patients with DL-AS [20]. *A* EcoRI (R1) genomic map of COL4A5 and COL4A6 5′ ends. The COL4A5 and COL4A6 exons are shown by striped and black boxes. respectively. *B* Extent of the deletions. Open boxes show the minimum extent of the deletions, black boxes represent the nondeleted regions, and hatched boxes represent ambiguities in mapping the deletion boundaries.

with AS but without leiomyomatosis and bearing a deletion of the 5′ part of COL4A5 have been shown to have larger deletion in COL4A6, 1 of them being deleted of the whole gene [21].

Different hypotheses can be proposed in order to explain these results:

(1) The DL process could be due to the absence of α6(IV) chain. It is now well known that interactions between cells and the extracellular matrix proteins, many of them being mediated by integrin family of cell surface receptors [24], play a vital role in control of cell behavior, growth and differentiation [25]. In the presence of antibodies directed against a β-integrin, chicken embryo myoblasts continue to replicate, fail to fuse, and have an abnormal phenotype [26]. Type IV collagens contain binding sites for several cell types including myocytes [27, 28]. Although all the studies have been performed with collagen sources rich in the ubiquitous α1 and α2 type IV chains, the α5 and α6 chains may have similar roles in cell-extracellular matrix interactions in tissues involved in AS-DL. The absence of the α6 or both α5 and α6 IV chains may disrupt these interactions and lead to benign smooth muscle cell proliferation. However, given the fact that 3 male patients with large deletions in COL4A6 are affected with DL, such hypothesis could only be retained if DL is considered as incompletely penetrant, or behaves with highly variable expression. This is unlikely given our recent study of 15 females in 6 DL-AS families, in which DL segregates as a completely penetrant dominant trait, and seems as severe in females as in males, contrary to the renal symptomatology which is generally mild and can even be totally absent in female [Dahan, in preparation].

(2) Alternatively, the DL process could be due to a mutation of a third gene lying within the second intron of COL4A6, behaving as a tumor or growth suppressor gene [29]. Such genes embedded within introns of other genes have been described in the NF1 gene [30] and in the human factor VIII gene [31]. However, it assumes that the complete deletion of this putative gene does not lead to any phenotype, since patients with deletion removing completely the second COL4A6 intron are not DL affected.

(3) Finally, DL could involve the presence of an abnormal truncated α6(IV) protein, translated from a truncated transcript. Indeed the presence of a COL4A6 mRNA has been shown in an esophageal tumor sample [21], suggesting the existence of a cryptic promoter located in the second COL4A6 intron. This hypothesis could explain both the dominant effect of the mutation and the lack of tumor in the 3 patients with larger deletions. However, it raises the question of a functional redundancy of the α6(IV) chain.

Whatever the gene involved in DL, these results demonstrate that the same gene, COL4A5, is involved in classical AS and in AS associated with DL. Furthermore, DL illustrates the crucial role of the extracellular matrix in cell differentiation and proliferation.

Diffuse Esophageal Leiomyomatosis without AS

Cases of diffuse esophageal leiomyomatosis without AS have been reported. Autosomal dominant inheritance [3, 4] or absence of renal symptoms in male patients over 60 [32] clearly exclude the involvement of COL4A5/COL4A6 in these cases. In sporadic cases, involvement of COL4A6 alone could be proposed, but the occurrence of the symptoms so late in life and the absence of cataracts in all cases makes this hypothesis unlikely. On the contrary, some cases of DL without AS, either isolated sporadic [32] or familial [33], or associated with genital involvement [34, 35] or cataract [36], could belong to the same entity as previously described, since all the affected patients are female and could have minor renal or hearing symptoms which have not been looked for, or even no symptom at all. Analysis of the DNA of such patients with COL4A5 and COL4A6 probes could confirm this hypothesis.

References

1   Heald J, Moussalli H, Hasleton PS: Diffuse leiomyomatosis of the oesophagus. Histopathology 1986;10:755–759.

2   Bourque MD, Spigland N, Bensoussan AL, Collin PP, Saguem MH, Brochu P, Blanchard H, Reinberg O: Esophageal leiomyoma in children: Two case reports and review of the literature. J Pediatr Surg 1989;85:303–305.

3   Marshall JB, Diaz-Arias AA, Bochna GS, Vogele KA: Achalasia due to diffuse esophageal leiomyomatosis and inherited as an autosomal dominant disorder. Report of a family study. Gastroenterology 1990;98:1358–1365.

4   Rosen RM: Familial multiple upper gastrointestinal leiomyoma. Gastroenterology 1990;85:303–305.

5   Garcia-Torres R, Guarner V: Leiomiomatosis del esophago, traqueo bronquial y genital asociada con nefropatia hereditaria tipo Alport: un nuevo sindrome. Rev Gastroenterol Méx 1983;48:163–170.

6   Garcia-Torres R, Orozco L: Alport-leiomyomatosis syndrome: An update. Am J Kidney Dis 1993; 22:641–648.

7   Roussel B, Birembaut P, Gaillard D, Puchelle JC, D'Albignac G, Pennaforte F, Fandre M: Léiomyomatose oesophagienne familiale associée à un syndrome d'Alport chez un garçon de 9 ans. Helv Paediatr Acta 1986;41:359–368.

8   Cochat P, Guibaud P, Garcia Torres R, Roussel B, Guarner V, Larbre F: Diffuse leiomyomatosis in Alport syndrome. J Pediatr 1988;113:339–343.

9   Leichter HE, Vargas J, Cohen AH, Ament M, Salusky IB: Alport's syndrome and achalasia. Pediatr Nephrol 1988;2:312–314.

10   Hasegawa S, Saitoh A, Imai A, Kikuta Y, Matsumoto Y, Ohi R: A case of non-Alport's hereditary nephritis with esophageal leiomyomatosis. Paediatr Nephrol 1989;3:C188.

11   Leborgne J, Le Neel JC, Heloury Y, Audoin AF, David A, Babut JM, Lenne Y: La léiomyomatose oesophagienne diffuse. A propos de 5 observations dont 2 cas familiaux. Chirurgie 1989;115:277–286.

12   McCartney PJ, McGuiness R: Alport's syndrome and the eye. Aust NZ J Ophthal 1989;17:165 168.

13   Witting BM, Treichel U, Kohler H, Rumpelt HJ, Meyer zum Buschenfelde KH: Leiomyomatose des Gastrointestinal- und Urogenitaltrakts in Kombination mit hereditärer Nephritis. Med Klin 1990;85:122–124.

14   Legius E, Proesmans W, Van Damme B, Geboes K, Lerut T, Eggermont E: Muscular hypertrophy of the oesophagus and 'Alport-like' glomerular lesions in a boy. Eur J Pediatr 1990;149:623–627.

15 Rabushka LS, Fishman EK, Kuhlman JE, Hruban RH: Diffuse esophageal leiomyomatosis in a patient with Alport syndrome: CT demonstration. Radiology 1991;179:176–178.

16 Lerone M, Dodero P, Romeo G, Martuciello G, Caffarena, Brisigotti M, Toma P, Taccone A, Silengo M: Leiomyomatosis of oesophagus, congenital cataracts and hematuria. Report of a case with rectal involvement. Pediatr Radiol 1991;21:578–579.

17 Lonsdale RN, Roberts PF, Vaughan R, Thiru S: Familial oesophageal leiomyomatosis and nephropathy. Histopathology 1992;20:127–133.

18 Bloch P, Quijada J: Léiomyomatose diffuse de l'oesophage. Analyse d'un cas et revue de la littérature. Gastroenterol Clin Biol 1992;16:890–893.

19 Antignac C, Zhou J, Sanak M, Cochat P, Roussel P, Deschênes G, Gros F, Knebelmann B, Hors-Cayla MC, Tryggvason K, Gubler MC: Alport syndrome and diffuse leiomyomatosis: Deletions in the COL4A5 collagen gene. Kidney Int 1992;42:1178–1183.

20 Zhou J, Mochizuki T, Smeets H, Antignac C, Laurila P, de Paepe A, Tryggvason K, Reeders ST: Deletion of the paired α5(IV) and α6(IV) collagen genes in inherited smooth muscle tumors. Science 1993;261:1167–1169.

21 Heidet L, Dahan K, Zhou J, Xu Z, Cochat P, Gould JDM, Leppig KA, Proesmans W, Guyot C, Guillot M, Roussel B, Tryggvason K, Grünfeld JP, Gubler MC, Antignac C: Deletions of both α5(IV) and α6(IV) collagen chains associated with smooth muscle tumours. Hum Mol Genet 1995; 4:99–108.

22 Kenney LJ: Giant intramural leiomyoma of esophagus. A case report. J Thorac Surg 1953;26: 93–100.

23 Johnston JB, Clagett OT, McDonald R: Smooth muscle tumors of the oesophagus. Thorax 1953; 8:251–265.

24 Hynes RO: Integrins: versatility, modulation, and signalling in cell adhesion. Cell 1992;69:11–25.

25 Madri JA, Masson MD: Extracellular matrix-cell interactions: Dynamic modulators of cell, tissue and organism structure and function. Lab Invest 1992;66:519–521.

26 Menko AS, Boettiger D: Occupation of the extracellular matrix receptor, integrin, is a control point for myogenic differentiation. Cell 1987;51:51–57.

27 Aumailley M, Timpl R: Attachment of cells to basement membranes. J Biol Cell 1986;103:1569–1575.

28 Vandenberg P, Kern A, Ries A, Luchenbill-Edds L, Mann K, Kühn K: Characterization of a type IV collagen major binding site with affinity to the α1β1 and α2β1 integrins. J Cell Biol 1991;113: 1475–1483.

29 Weinberg RA: Tumor suppressor genes. Science 1991;254:1138–1146.

30 Viskochil D, Cawthon R, O'Connell P, Xu G, Stevens J, Culver M, Carey J, White R: The oligodendrocyte-myelin glycoprotein is embedded within the neurofibromatosis type 1 gene. Mol Cell Biol 1991; 11:906–912.

31 Levinson B. Kenwrick S, Lakich D, Hammonds G, Gitschier J: A transcribed gene in an intron of the human factor VIII gene. Genomics 1990;7:1–11.

32 Fernandes JP, Mascarenhas MJ, da Costa JC, Correia JP: Diffuse leiomyomatosis of the esophagus. A case report and review of the literature. Dig Dis 1975;20:684–690.

33 Wachsmuth W: Über hereditäres Vorkommen zirkulär wachsender Leiomyome des oesophagus. Chirurg 1959;30:145–149.

34 Schapiro RL, Sandrock AR: Esophagogastric and vulvar leiomyomatosis: A new radiologic syndrome. J Can Assoc Radiol 1973;23:184–187.

35 Wahlen T, Astedt B: Familial occurrence of coexisting leiomyoma of the vulva and oesophagus. Acta Obstet Gynecol Scand 1965;44:197–203.

36 Blank E, Michael TD: Muscular hypertrophy of the esophagus: Report of a case with involvement of the entire esophagus. Pediatrics 1963;32:595–598.

37 Zhou J, Leinonen A, Tryggvason K: Structure of the type IV collagen COL4A5 gene. J Biol Chem 1994;269:6608–6614.

Dr. Corinne Antignac, INSERM U423, Tour Lavoisier, 6 étage, Hôpital Necker – Enfants Malades, F–75743 Paris Cedex 15 (France)

Tryggvason K (ed): Molecular Pathology and Genetics of Alport Syndrome.
Contrib Nephrol. Basel, Karger, 1996, vol 117, pp 183–197

New Approaches to the DNA Diagnosis of Alport Syndrome

Alessandra Renieri, Mario De Marchi

Medical Genetics Unit, Department of Molecular Biology, University of Siena, and
Department of Clinical and Biological Sciences, University of Torino,
Orbassano, Italy

The Rationale of Mutation Analyses

When a new 'disease gene' is first identified in the genome, pinpointing of mutated alleles is a necessary exercise to support its implication as responsible, or at least correspondent, in the etiology of the disease. This is particularly critical when the gene is cloned through the positional and 'candidate gene' approaches, as has happened for COL4A5 in Alport syndrome (AS) [1–3]. In addition, direct identification of mutations provides a significant, though indirect, contribution to the understanding of the pathogenic mechanisms of a disease, and can even become useful as a diagnostic tool and for epidemiological studies when extended to larger patient series. However, while a few pathogenic mutations give cogent evidence for the first goals, the sensitivity, cost and performance of the screening procedure are critical for the subsequent diagnostic application. Thus, different uses have different requirements, and prompt the use of different mutation analysis methods (table 1).

Principles and Levels of Mutation Detection

The choice of a method for detection of DNA sequence variants should also take into account the structure and expression pattern of the gene to be studied. The molecular diagnosis of AS is a veritable challenge for molecular biologists because of its locus and allele heterogeneity (chapters 1, 2, 5), so that (1) it may be necessary to screen more than one gene, (2) for mutations

Table 1. Possible uses of mutation identification

Aim of the study	Technical requirements
Support of disease gene identification	Rapidity and simplicity
Direct molecular diagnosis (vs. indirect through linkage)	High sensitivity
Clues to the mutagenic and pathogenetic mechanisms	Detailed characterization
Mutation/phenotype correlations	Large series of patients
Population genetics	Large series of patients

differing from one to another (chapter 8). Moreover, (3) major gene rearrangements, easily detectable through standard Southern blotting, account for only 5–10% of X-linked AS [4–9], whereas most mutations are small deletions/ insertions and single-base changes in the coding region and at exon-intron splice sites [10–15]. Thus, molecular diagnosis of AS cannot use techniques for identifying a known set of major disease alleles, such as hybridization to allele-specific oligonucleotides (ASO) [16], but must rely on scanning methods able to pinpoint the presence of unknown mutations in any position of the gene. Lastly, (4) the COL4 genes to be screened rank among the largest known for exon number, and (5) the target tissues are relatively inaccessible [17].

Genomic DNA vs. mRNA as Starting Material

Were it not for this last problem, mRNA would probably be the most appropriate choice for molecular diagnosis, in particular for X-linked genes, in which the pathologic transcript changes in males are not obscured by the normal ones. There are, in fact, specialized protocols that exploit the so-called 'illegitimate transcription' [18]. Even starting from cells in which the gene is expressed at very low level, such as lymphocytes or lymphoblasts, it is possible to retro-transcribe the mRNA and then obtain enough material for mutation scanning by two stages, nested PCR (RT-PCR). By choosing appropriate primers, the entire mRNA can thus be analyzed into a few segments, each encompassing several exons. This procedure can both directly detect mutations in the coding sequences, as well as changes in transcript level or structure, e.g. abnormal splice products, caused by mutations in intronic sequences flanking the exons [14, 19], and has even been proved able to indirectly identity mutations in the middle of introns, that would otherwise have been missed [20]. The occurrence in COL4 genes of alternative, tissue specific splice patterns is a potential limitation of the technique [21, 22]. In practice, it has proved difficult to obtain suitable mRNA from all patients, and DNA-based techniques are most commonly chosen.

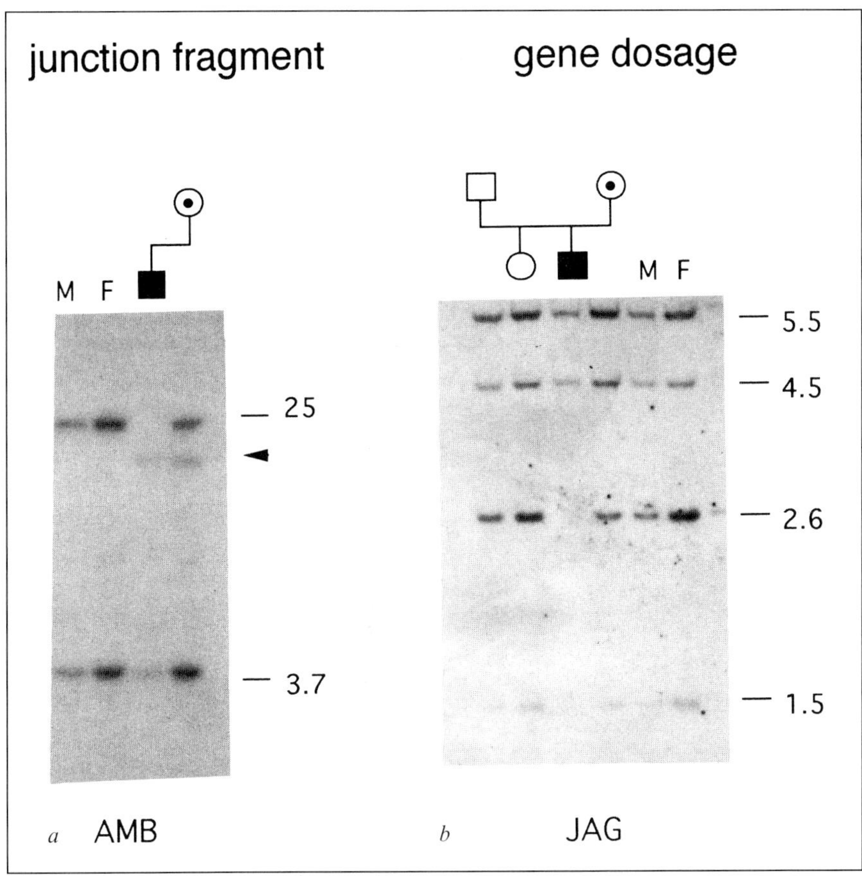

Fig. 1. Southern blotting pattern of two intragenic COL4A5 deletions. *a* In family AMB, hybridization of TaqI filter with probe PF17A shows a junction fragment in both the male patient and his heterozygous mother (arrowhead). *b* No junction fragment was detected in family JAG, and the mother's carrier status could be verified by densitometry: hybridization of an EcoRI filter with probe PF6 shows half dose of the 2.6- and 1.5-kb bands compared to the 5.5- and 4.5-kb ones. Product length is indicated in kb. M = Male; F = female.

Strategies for Identifying Mutations at the DNA Level

Southern blotting analysis is a classic first step in the mutation diagnosis of newly-found genes, as it allows an early jump to the genome as soon as cDNA probes are available. In X-linked diseases, large deletions are easily detected by the absence of bands in males. Diagnosis in female patients relies on gene dosage evaluation, unless a junction fragment is identified with the right enzyme (fig. 1). In our experience, EcoRI junction fragments are found in 70% of the COL4A5 rearrangements, and this rises to 80% if more enzymes

Table 2. Comparison of different mutation scanning methods

	Simplicity	Time-saving	Sensitivity	Cost
DGGE	+ +	+ + +	+ +	+ +
CMC	+	+ +	+ + +	+ +
PTT	+	+	+	+ +
SSCP	+ + +	+ + +	+ +	+
Automated sequencing	+ +	+ + +	+ + +	+ + +

are tested. Upgrading of this exploratory analysis to fine characterization of the rearrangements demands many additional efforts with different enzyme/probe combinations. A detailed map of both the normal and the rearranged genes will, for example, allow definition of the breakpoints, which could provide clues on the mutagenic mechanism [23], or dissection of a complex phenotype into a contiguous gene syndrome by deletion mapping (see the example of AS + leiomyomatosis, chapter 9). In this respect, Southern blotting, even though it is still widely used, may already be obsolete [24], and other approaches may often be more useful. In cases with very large deletions, PFGE analysis can help to estimate the extension of the rearrangement [25]. When the normal gene structure is known, PCR of single exons offers an alternative method for both screening and characterizing the deletions. Characterization of the exon-intron structure of COL4A5 [17], has made it straightforward to analyze the entire coding sequence in segments. Different laboratories now follow the common strategy of amplifying all exons using primers tailored on the flanking intronic sequences; many PCR products are thus obtained, each containing one exon plus a short stretch of intron sequence, except for fragments that contain pairs of close exons, e.g. exons 11–12 and 14–15. In male patients, agarose gel electrophoresis will reveal deletions as absences of the corresponding amplification products. Time and reagents can be saved by simultaneously amplifying several segments throughout the gene (multiplex PCR). This approach is widely used and has substituted Southern blotting in the diagnosis of Duchenne muscular dystrophy [26]. The PCR products can then be used in mutation scanning methods.

Available Methods for Mutation Identification

The introduction of scanning methods for detection of point mutations, followed by direct sequencing, was a revolution in this field. These methods now include: chemical mismatch cleavage (CMC), denaturing gradient gel

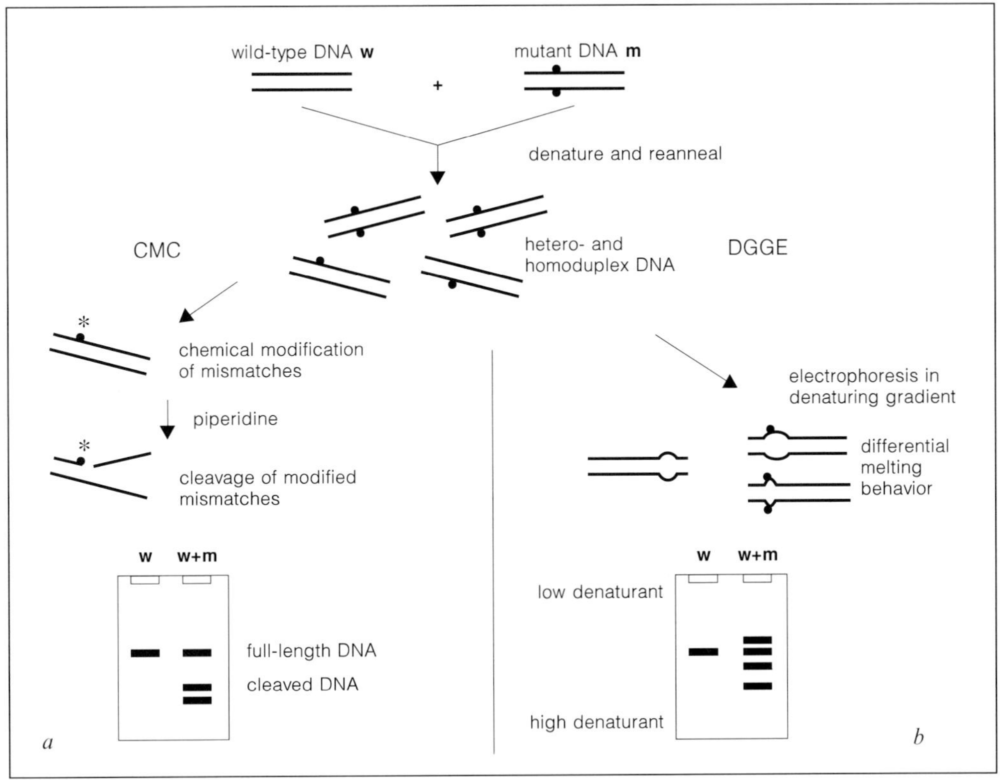

Fig. 2. Principles for detecting heteroduplex DNA molecules by (*a*) CMC and (*b*) DGGE. In both methods, analysis of hemizygous male samples requires mixing of the patient's sample with normal reference DNA.

electrophoresis (DGGE), single-strand conformation polymorphism (SSCP), and the protein truncation test (PTT). Their main features are shown in table 2. Several of these methods rest on a common basic principle, i.e. the formation of mismatched DNA-DNA heteroduplexes between the normal and the mutated strands (fig. 2). Heteroduplexes can then be recognized either through their distinctive chemical or enzymatic reactivity, their melting behavior or their differential migration in gels [27]. A different principle is at work in SSCP, in which single-stranded molecules are compared (fig. 3). Other methods, such as PTT, analyze peptide products and only detect stop mutations.

CMC was first described by Cotton et al. [28]. It enables the identification of single-base changes in heteroduplexes formed between radiolabelled wild-type DNA and nonlabelled mutant DNA. Unpaired bases are chemically modified with either osmium tetroxide, which modifies thymidine (T), or

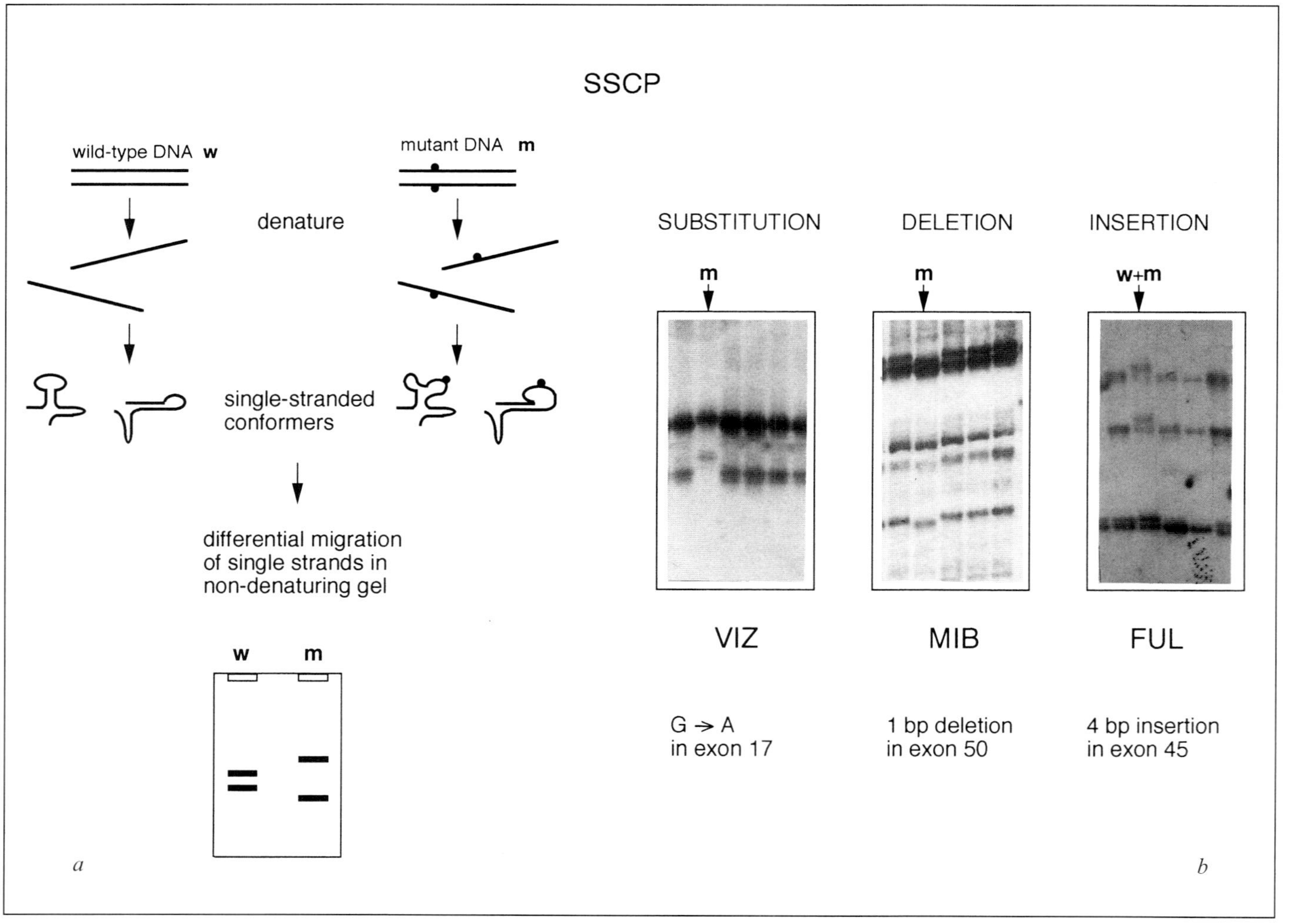

SSCP
wild-type DNA w
mutant DNA m
denature
single-stranded conformers
differential migration of single strands in non-denaturing gel
w
m
SUBSTITUTION
m
VIZ
G → A
in exon 17
DELETION
m
MIB
1 bp deletion
in exon 50
INSERTION
w+m
FUL
4 bp insertion
in exon 45
a
b

hydroxylamine, which reacts with cytosine (C). In a subsequent reaction, the heteroduplex molecule is then specifically cleaved by piperidine at the chemical modification site; the cleaved products can be easily visualized by denaturing polyacrylamide gel electrophoresis (PAGE) and autoradiography (fig. 2a). Fragment length indicates the exact distance of the cleaved mismatch from the radiolabelled end. By labelling both sense and antisense strands of wild-type DNA, virtually all possible mismatches can be detected. CMC has the lowest size constraint of all scanning methods, up to 1.7 kb, and thus permits a most efficient screening. An application of this technique to genomic DNA for mutation scanning of the COL4A5 has been reported [29]. Technical improvements currently under scrutiny include nonisotopic techniques [30] and enzymatic recognition of the mismatch with bacterial proteins of the repair machinery, e.g. mutS [31], or with phage T4 resolvase [32].

DGGE discriminates mismatched DNA heteroduplexes on the basis of their melting properties in a gel [33]. Double-stranded DNA is subjected to electrophoresis through a denaturant gradient. As DNA migrates into the gel, the denaturing agent concentration increases, and strand separation starts in the region of the heteroduplex molecule with the lowest melting point. As partially branched DNA migrates slower than its double-stranded equivalent, heteroduplexes of equal size but different melting points will be resolved in the gel (fig. 2b). Even single-base changes will show a shifted migration, though those falling in a domain with the highest melting point, which stay double-stranded throughout the run, remain undetected. This limitation can be over-come by attaching an artificial domain with an extremely high melting point to the target molecule, in practice by joining a GC clamp to the primers used for amplification [34]. In principle, this makes any sequence difference detectable in fragments of up to 600 bp of length. Even so, there are two drawbacks: (a) the need to accurately define the melting profile of each new sequence in order to adjust the denaturant gradient and (b) the additional cost of synthesizing longer primers for each fragment to be analyzed. Several mutations on COL4A5 gene have been detected by this method [10, 35, 36].

PTT provides a rapid and sensitive way to screen large coding sequences specifically for mutations causing premature stop codons [37]. Generally, RNA is isolated from fresh blood or tissue samples, reverse-transcribed and amplified by PCR. When analyzing very large exons, genomic DNA can also be used

Fig. 3. SSCP analysis. *a* Basic principle of the method. *b* Patterns of three different kinds of small mutations in the COL4A5 gene. VIZ [11] and MIB [13] are males, FUL [45] a female. Each single strand can originate multiple conformers. In all cases run conditions were: 6% acrylamide with 10% glycerol, run at 4 °C, 5 W/40 cm, overnight.

as template for PCR. Deletions, and, by RT-PCR, splice site mutations as well, are detected at this step as shorter PCR products. A second round of PCR is then performed using a modified forward primer containing a T7 promoter with an in-frame translation initiation codon, and these PCR products are used as in vitro templates for a coupled transcription/translation assay, in the presence of labelled emino acids. The appearance of a truncated peptide in SDS-PAGE pinpoints the presence of translation termination mutations. This method is focused on the coding part and insensitive to silent mutations. It is useful for scanning genes with enormous size and large number of exons, and is particularly suited for the diagnosis of those diseases in which most mutations are premature translation termination, such as DMD, APC, NF2, etc. However, as it also ignores missense changes and other nontruncating mutations, the negative fragments should be further investigated by SSCP or any of the other above described techniques. Indeed, PTT does not seem elective for AS, where truncating mutations constitute only 8/20 of all the small mutations reported (chapter 8).

The simplest way for detection of point mutations is offered by SSCP [38]. In this method, the DNA sample is denatured and subjected to PAGE under nondenaturing conditions. Thus, both single strands of DNA take a folded conformation, stabilized by intra-strand interactions. Nucleotide sequence changes cause a different conformation and therefore a different mobility (fig. 3). While most mutations can be detected under different run conditions, some have been reported to depend on a particular temperature or the presence or absence of glycerol [38]. This would imply the need to repeat the test in two or more ways to maximize its sensitivity. The original isotopic technique requiring autoradiography has been modified into nonisotopic procedures by using either digoxigenin labelling [13], or silver as well as ethidium staining of the gel [39, 40] (fig. 4). Most mutation analyses on the COL4A5 gene have been performed by this technique, either isotopic [7, 41–43] or nonisotopic [13, 15, 44–46].

Automated Sequencing
Direct sequencing can, in one sense, be viewed as a crude method, since all fragments of the gene are analyzed, without prior information about the presence of a mutation. Screening of the COL4A5 gene has been attempted manually [12], but it can be significantly improved by automation. In the current technology, the PCR product is affinity-purified and used as the template in sequencing Sanger reactions, starting from an internal oligonucleotide primer fluorescently labelled at the 5′ end [47]. The band pattern is then read after or even during the electrophoretic run, using a computerized laser apparatus. In the near future, automation will be extended to all steps, from

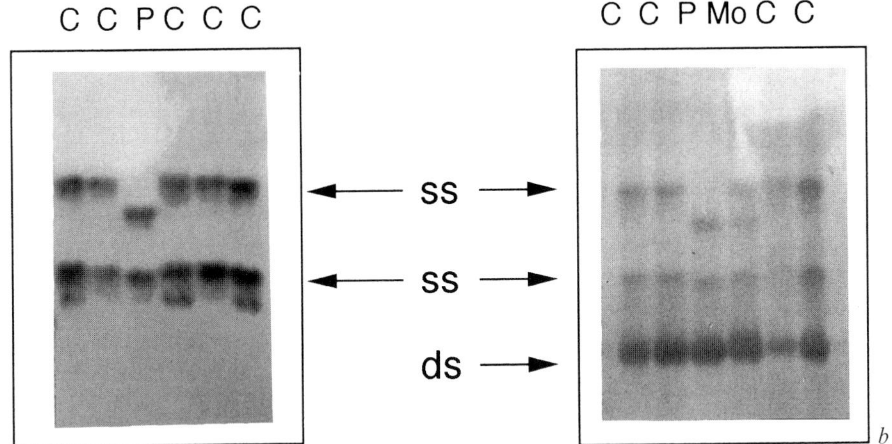

Fig. 4. Nonisotopic SSCP of exon 39 of the COL4A5 gene, detected by (*a*) hybridization with digoxigenin-labeled probe and (*b*) silver staining. C = Control; P = patient; Mo = mother. Run condition as figure 3. The altered bands bear two silent point mutations [9]. Hybridization preferentially detects single-strand conformers (ss), whereas silver staining stochiometrically reveals both single-strand and homoduplex double-stranded molecules (ds).

set-up of enzymatic reactions to sample purification and loading. Hopefully, the resulting increased productivity will reduce the current high cost per nucleotide. Thanks to its sensitivity, direct sequence analysis may itself even do away with the need of mutation scanning.

A Protocol for the Molecular Diagnosis of AS

Mutation scanning has until now been mainly driven by its intrinsic research component, and experience of its application for routine diagnostic application is still limited. An attempt will be made to summarize some reflections in this perspective. The first step in an AS diagnostic protocol is accurate clinical definition of the phenotype plus a formal genetics analysis of the family (fig. 5). This may itself be sufficient to discriminate families with X-linked inheritance, requiring COL4A5 analysis, from those with autosomal inheritance, where analysis of the COL4A3/COL4A4 genes is needed. Sporadic cases should be screened for both genes, because they can be either de novo COL4A5 mutations or autosomal recessive forms. For the prevalent X-linked cases, if the proband's clinical diagnosis is clear-cut, linkage analysis is the method of choice for preclinical and carrier diagnosis in relatives, on account

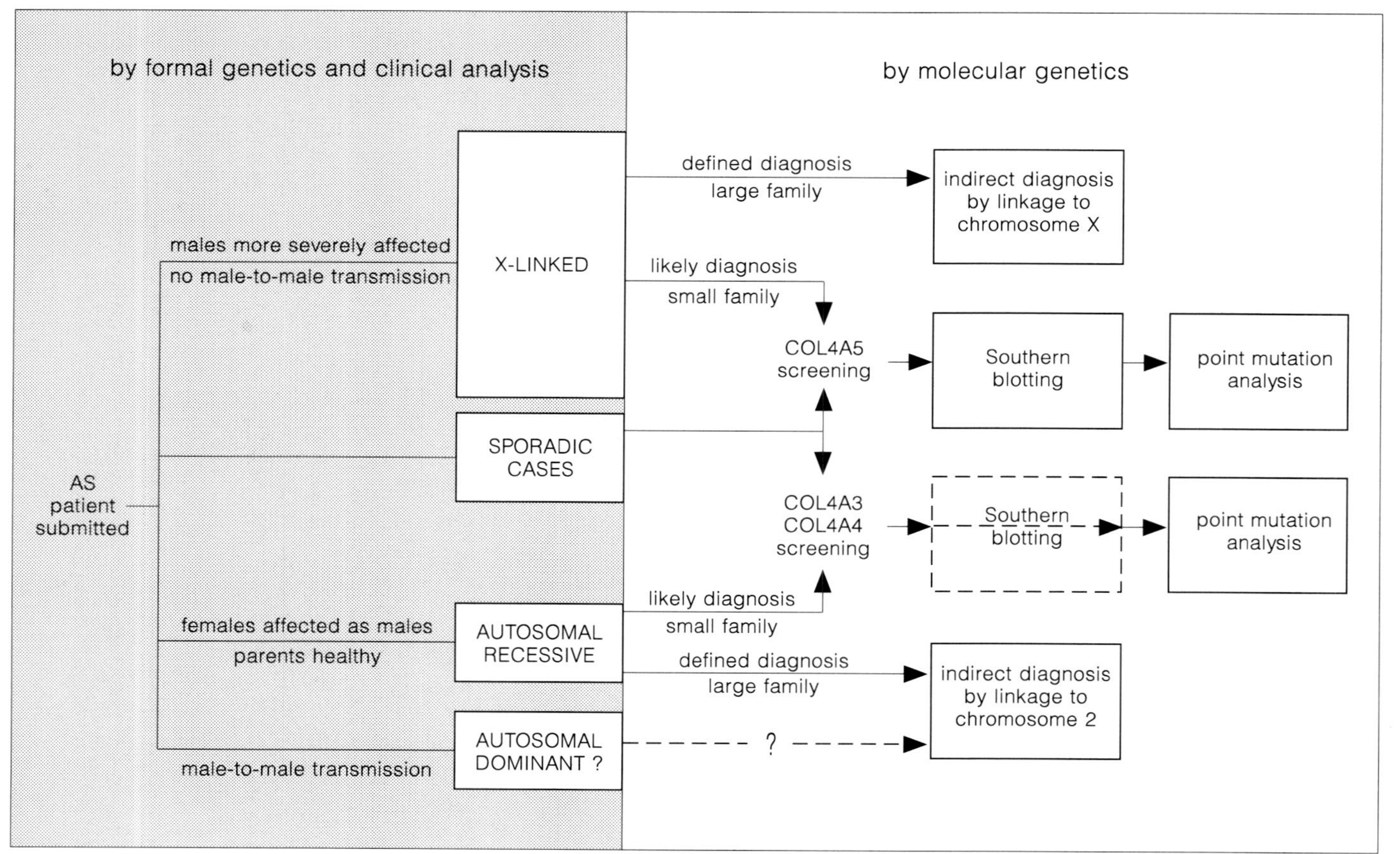

by formal genetics and clinical analysis
by molecular genetics
AS patient submitted
males more severely affected
no male-to-male transmission
X-LINKED
defined diagnosis large family
indirect diagnosis by linkage to chromosome X
likely diagnosis small family
COL4A5 screening
Southern blotting
point mutation analysis
SPORADIC CASES
COL4A3 COL4A4 screening
Southern blotting
point mutation analysis
females affected as males
parents healthy
AUTOSOMAL RECESSIVE
likely diagnosis small family
defined diagnosis large family
indirect diagnosis by linkage to chromosome 2
male-to-male transmission
AUTOSOMAL DOMINANT ?
?

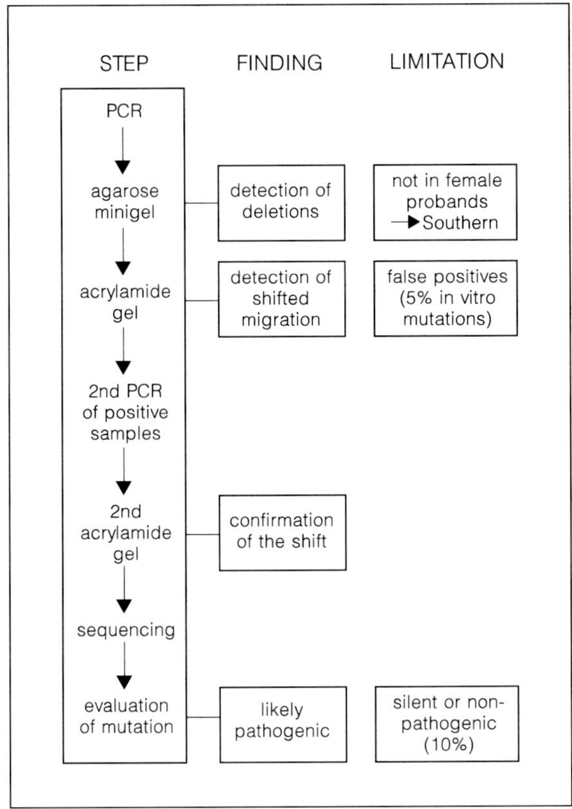

Fig. 6. Outline of the suggested protocol for molecular diagnosis of AS.

of its rapidity and efficiency. If diagnosis is uncertain, the proband DNA (preferably male) should be screened for mutations. With a large acrylamide gel apparatus (60 wells), screening of 10 patients for all the COL4A5 exons with one SSCP condition should take about 2 weeks. One or two additional workdays are necessary for preparation of the DNA, sequencing and evaluation of the mutation. Note that this suggested protocol (fig. 6) does not include Southern blotting, except for female probands. After PCR, an agarose minigel is useful both to detect deletions and check the amount and purity of the products before their run in the SSCP native polyacrylamide gel. Every time a shift is obtained, its true nature should be tested on the product of a repeated PCR. In our experience, in fact, about 5% of shifts were false positives, i.e.

Fig. 5. Strategies for molecular diagnosis of AS.

mutations arose in vitro during the PCR. Of the confirmed shifts, about
10% were silent mutations or nonpathogenic amino acid substitutions. Null
mutations are likely to be pathogenic. This conclusion is more questionable
in case of missense changes, and, as discussed in chapter 8 by Tryggvason,
requires evolutionary comparison of the amino acid sequence in related genes
and different species. However, even changes affecting evolutionary conserved
residues may not segregate with the disease [48].

By applying the protocol described to 80% of exons, and extrapolating
for the entire gene, we may now expect to find a mutation in 30% of AS
patients. Even if a fraction of 5–7% with large deletions is added, diagnosis
is made in our hands in less than 40% of cases. This finding supports the
usefulness of linkage analysis whenever possible, and should be considered
when planning a diagnostic protocol. Possible explanations of this low detection rate are: heterogeneity of the disease, phenocopies, mutations in noncoding
sequences and the limited sensitivity of SSCP. Even if the effectiveness of
SSCP is assumed to be lower in comparison with DDGE, it has been estimated
to be at least 80% [49] and even better for short PCR products. Actually, the
extrapolated detection rate in our series was as high as 70% in typical cases
of the juvenile-type and about 50% in typical adult-type AS [50]. Nevertheless,
analysis of large genes is still demanding. All presently available methods are
time-consuming, inefficient or expensive, and further developments are needed
before molecular scanning can become the routine approach to the diagnosis
of AS.

Acknowledgments

The work of our laboratories was supported by Telethon-Italy (Grant No. E.134), and
Fondazione Rulfo per la Genetica Medica. The contribution of many patients, relatives and
colleagues of the Italian Multicenter Study on Alport syndrome is acknowledged. We are
indebed to Prof. K. Tryggvason for unpublished information on the COL4A5 gene structure,
and to Drs. M. Bruttini and L. Galli for dedicated collaboration.

References

1   Hostikka SL, Eddy M, Byers MG, Höyhtyä M, Shows TB, Tryggvason K: Identification of a
    distinct type IV collagen α chain with restricted tissue distribution and assignment of its gene to
    the locus of X chromosome-linked Alport syndrome. Proc Natl Acad Sci USA 1990;87:1606–
    1610.
2   Myers JC, Jones TA, Pohjolainen ER, Kadri AS, Goddard AD, Sheer D, Solomon E, Pihlajaniemi
    T: Molecular cloning of α5(IV) collagen and assignment of the gene to the region of X chromosome
    containing the Alport syndrome locus. Am J Hum Genet 1990;46:1024–1033.

3   Barker DF, Hostikka SL, Zhou J, Chow LT, Oliphant AR, Gerken SC, Gregory MC, Skolnick MH, Atkin CL, Tryggvason K: Identification of mutations in the COL4A5 collagen gene in Alport syndrome. Science 1990;248:1224–1227.

4   Boye E, Vetrie D, Flinter F, Buckle B, Pihlajaniemi T, Hamalainen ER, Myers JC, Bobrow M, Harris A: Major rearrangements in the α5(IV) collagene gene in three patients with Alport syndrome. Genomics 1991;11:1125–1132.

5   M'Rad R, Sanak M, Deschenes G, Zhou J, Bonaiti-Pellie C, Holvoet-Vermaunt L, Heuertz S, Gübler MC, Broyer M, Grünfeld JP, Tryggvason K, Hors-Cayla MC: Alport syndrome: A genetic study of 31 families. Hum Genet 1992;90:420–426.

6   Netzer KO, Reeders L, Zhou J, Pullig O, Tryggvason K, Weber M: Deletions of the COL4A5 gene in patients with Alport syndrome. Kidney Int 1992;42:1336–1344.

7   Smeets HJM, Melenhorst JJ, Lemmink HH, Schröder CH, Nelen MR, Zhou J, Hostikka SL, Tryggvason K, Ropers HH, Jansweijer MCE, Monnens LAH, Brunner HG, van Oost BA: Different mutations in the COL4A5 gene in two patients with different features of Alport syndrome. Kidney Int 1992;42:83–88.

8   Antignac C, Knebelmann B, Drouot L, Gros F, Deschenes G, Hors-Cayla MG, Zhou J, Tryggvason K, Grünfeld JP, Brioyer M, Gübler MC: Deletions in the COL4A5 collagen gene in X-linked Alport syndrome. J Clin Invest 1994;93:1195–1207.

9   Renieri A, Galli L, Grillo A, Bruttini M, Neri T, Zanelli P, Rizzoni G, Massella L, Sessa A, Meroni M, Peratoner L, Riegler P, Scolari F, Mileti M, Giani M, Cossu M, Savi M, Ballabio A, De Marchi M: Major COL4A5 gene rearrangements in patients with juvenile-type Alport syndrome. Am J Med Genet 1995; in press.

10   Zhou J, Hertz JM, Tryggvason K: Mutation in the α5(IV) collagen chain in juvenile-onset Alport syndrome without hearing loss or ocular lesions: Detection by denaturing gradient gel electrophoresis of a PCR product. Am J Hum Genet 1992;40:1291–1300.

11   Renieri A, Seri M, Myers JC, Pihlajaniemi T, Massella L, Rizzoni G, De Marchi M: De novo mutation in the COL4A5 gene converting glycine 325 to glutamic acid in Alport syndrome. Hum Mol Genet 1992;1:127–129.

12   Nomura S, Osawa G, Sai T, Harano T, Harano K: A splicing mutation in the α5(IV) collagene gene of a family with Alport syndrome. Kidney Int 1993;43:1116–1124.

13   Renieri A, Seri M, Galli L, Cosci P, Imbasciati E, Massella P, Rizzoni GF, Restagno G, Carbonara AO, Stramignoni E, Basolo B, Piccoli G, De Marchi M: Small frameshift deletions within the COL4A5 gene in juvenile-onset Alport syndrome. Hum Genet 1993;92:417–420.

14   Lemmink HH, Kluijtmans AJ, Brunner HG, Schröder CH, Knebelmann B, Jelinkova E, van Oost BA, Monnens LAH, Smeets JM: Aberrant splicing of the COL4A5 gene in patients with Alport syndrome. Hum Mol Genet 1994;3:317–322.

15   Renieri A, Meroni M, Sessa A, Battini G, Serbelloni P, Torri-Tarelli L, Seri M, Galli L, De Marchi M: Variability of clinical phenotype in a large Alport family with gly 1143 ser change of collagen α5(IV) chain. Nephron 1994;67:444–449.

16   Chehab FF, Wall J: Detection of multiple cystic fibrosis mutations by reverse dot blot hybridisation: A technology for carrier screening. Hum Genet 1992;89:163–68.

17   Zhou J, Leinonen A, Tryggvason K: Structure of the human type IV collagen COL4A5 gene. J Biol Chem 1994;269:6608–6614.

18   Chelly J, Concordet JP, Kaplan JC, Kahn JC: An illegitimate transcription: Transcription of any gene in any cell type. Proc Natl Acad Sci USA 1989;86:2617–2621.

19   Knebelmann B, Deschenes G, Gros F, Hors MC, Grunfeld JP, Tryggvason K, Gubler MC, Antignac C: Substitution of arginine for glycine 325 in the collagen α5(IV) chain associated with X-linked Alport syndrome: Characterisation of the mutation by direct sequencing of PCR-amplified lymphoblast cDNA fragments. Am J Hum Genet 1992;51:135–142.

20   Knebelmann B, Forestier L, Drouot L, Quinones S, Chuet C, Benessy F, Saus J, Antignac C: Splice-mediated insertion of an Alu sequence in the COL4A3 mRNA causing autosomal recessive Alport syndrome. Hum Mol Genet 1995;4:675–679.

21   Bernal D, Quinones S, Saus J: The human mRNA encoding the Goodpasture antigen is alternatively spliced. J Biol Chem 1993;268:12090–12094.

22 Guo C, Van Damme B, Van Damme-Lombaerts R, Van den Berghe H, Cassiman JJ, Marynen P: Differential splicing of COL4A5 mRNA in kidney and white blood cells: A complex mutation in the COL4A5 gene of an Alport patient deletes the NC1 domain, Kidney Int 1993;44:1316–1321.

23 Cooper DN, Krawczak M: Gene deletions; in Human Gene Mutation. Oxford, Bios Sci Publ Ltd, 1993, pp 163–176.

24 Williamson B: The end of an era. Hum Mol Genet 1995;4:149–151.

25 Vetrie D, Boye E, Flinter F, Bobrow M, Harris A: DNA rearrangements in the α5(IV) collagen gene (COL4A5) of individuals with Alport syndrome: Further refinement using pulsed-field gel electrophoresis. Genomics 1992;14:624–633.

26 Chamberlain JS, Gibbs RA, Ranier JE, Nguyen PN, Caskey CT: Deletion screening of the Duchenne muscular dystrophy locus via multiplex DNA amplification. Nucleic Acids Res 1988;16:11141–11156.

27 Ganguly A, Rock MJ, Prockop DJ: Conformation-sensitive gel electrophoresis for rapid detection of single-base differences in double-stranded PCR products and DNA fragments: Evidence for solvent-induced bends in DNA heteroduplexes. Proc Natl Acad Sci USA 1993;90:10325–10329.

28 Cotton RG, Rodrigues, NR, Campbell RD: Reactivity of cytosine and thymine in single base-pair mismatches with hydroxylamine and osmium tetroxide and its application to the study of mutations. Proc Natl Acad Sci USA 1988;85:4397–4401.

29 Boye E, Flinter FA, Bobrow M, Harris A: An 8 bp deletion in exon 51 of the COL4A5 gene of an Alport syndrome patient. Hum Mol Genet 1993;2:595–596.

30 Verpy E, Biasotto M, Meo T, Tosi M: Efficient detection of point mutations on color-coded strands of target DNA. Proc Natl Acad Sci USA 1994;91:1873–1877.

31 Jiricny J, Su S, Wood SG, Modrich P: Mismatch containing oligonucleotides bound by the *E coli* mutS-encoded protein. Nucleic Acids Res 1988;16:7843–7853.

32 Mashall RD, Koontz J, Sklar J: Detection of mutations by cleavage of DNA heteroduplex with bacteriophage resolvases. Nat Genet 1995;9:177–183.

33 Myers RM, Lumelsky N, Lerman LS, Maniatis T: Detection of single-base substitutions in total genomic DNA. Nature 1985;313:495–498.

34 Scheffield VC, Cox DR, Lerman LS, Myers RA: Attachment of a 40-base-pair G + C rich sequence (GC clamp) to genomic DNA by the polymerase chain reaction results in improved detection of single-base changes. Proc Natl Acad Sci USA 1989;86:232–236.

35 Netzer KO, Pullig O, Frei U, Zhou J, Tryggvason K, Weber M: COL4A5 splice site mutation and α5(IV) mRNA Alport syndrome. Kidney Int 1993;43:486–492.

36 Zhou J, Gregory MC, Hertz JM, Barker DF, Atkin C, Spencer E, Tryggvason K: Mutations in the codon for a conserved arginine-1563 in the COL4A5 collagen gene in Alport syndrome. Kidney Int 1993;43:722–729.

37 Van der Luijt R, Khan PM, Vasen H, van Leeuwen C, Tops C, Roest P, den Dunnen J, Fodde R: Rapid detection of translation-terminating mutations at the adenomatous polyposis coli gene by direct protein truncation test. Genomics 1994;20:1–4.

38 Orita M, Suzuki Y, Sekiya T, Hayashi R: Rapid and sensitive detection of point mutations and DNA polymorphisms using the polymerase chain reaction. Genomics 1993;5:874–879.

39 Dockhorn-Dworniczak B, Dworniczak B, Brömmelkamp L, Bülles J, Horst J, Böcker WW: Non-isotopic detection of single-strand conformation polymorphism (PCR-SSCP): A rapid and sensitive technique in diagnosis of phenylketonuria. Nucleic Acids Res 1991;19:2500.

40 Yap EPH, McGee JO'D: Non-isotopic SSCP detectionin PCR products by ethidium bromide staining. Trends Genet 1992;8:49.

41 Lemmink HH, Schröder CH, Brunner HG, Nelen MR, Zhou J, Tryggvason K, Haagsma-Shouten WAG, Roodvoets AP, Rasher W, van Oost BA, Smeets JM: Identification of four novel mutations in the COL4A5 gene of patients with Alport syndrome. Genomics 1993;17:485–489.

42 Nakazato H, Hattori S, Matsuura T, Koitabashi Y, Endo F, Matsuda I: Identification of a single-base insertion in the COL4A5 gene in Alport syndrome. Kidney Int 1993;44:1091–1096.

43 Boye E, Flinter FA, Zhou J, Tryggvason K, Bobrow M, Harris A: Detection of twelve novel mutations in the collagenous domain of the COL4A5 gene of Alport syndrome patients. Hum Mutat 1995;5:197–204.

44 Peissel B, Rossetti S, Renieri A, Galli L, De Marchi M, Battini G, Meroni M, Sessa A, Schiavano S, Pignatti PF, Turco AE: A novel frameshift deletion in type IV collagen a5 gene in a juvenile Alport syndrome patient: An adenine deletion (2940/2943 del A) in exon 34 of COL4A5. Hum Mutat 1994;3:386–390.

45 Massella L, Rizzoni GF, De Blasis R, Barsotti P, Faraggiana T, Renieri A, Seri M, Galli L, De Marchi M: De-novo COL4A5 gene mutations in Alport's syndrome. Nephrol Dial Transplant 1994; 9:1408–1411.

46 Renieri A, Galli L, De Marchi M, Lupo A, Li Volti S, Mollica F, Maschio G, Peissel B, Rossetti S, Pignatti GF, Turco AE: Single base pair deletions in exons 39 and 42 of the COL4A5 gene in Alport syndrome. Hum Mol Genet 1994;3:201–202.

47 Saarinen N, Zhou J, Leinonen A, Barker D, Gregory M, Atkin C, Netzer K, Weber M, Reeders S, Hertz JM, Shows T, Springate J, Gronhagen-Riska C, Neumann H, Trembath R, Tryggvason K: Large-scale mutation search in Alport syndrome. Third International Workshop on Alport Syndrome, Erlangen-Nürnberg 1994.

48 Lemmink HH, Mochizuki TJ, van den Heuvel LP, Schröder CH, Barrientos A, Monnens LAH, van Oost BA, Brunner HG, Reeders ST, Smeets HJL: Mutations in the type IV collagen α3(IV) gene in autosomal recessive Alport syndrome. Hum Mol Genet 1994;3:1209–1273.

49 Grompe M: The rapid detection of unknown mutations in nucleic acids. Nat Genet 1995;5:111–117.

50 Renieri A, Galli L, Bruttini M, Neri T, Zanelli P, Rossetti S, Turco A, Rizzoni GF, Massella L, Sessa A, Meroni M, Monga G, Mazzucco G, Barsotti P, Ballabio A, De Marchi M: Molecular genetics of Alport syndrome: An update of the Italian study. Third International Workshop on Alport Syndrome, Erlangen-Nürnberg 1994.

Prof. Mario De Marchi, Department of Clinical and Biological Sciences, University of Torino, Ospedale San Luigi, I–10043 Orbassano (Italy)